Introduction to Autodesk Land Desktop 2007
and
Civil Design 2007

Geoffrey J. Coleman, PE
Santa Rosa Junior College
Applied Technology Department

ISBN: 978-1-58503-354-6

PUBLICATIONS

Schroff Development Corporation

www.schroff.com
www.schroff-europe.com

Disclaimer

This manual has been prepared as a supplement to the CEST85 course taught at Santa Rosa Junior College. Users are required to have a superior knowledge of AutoCAD and a general understanding of Civil Engineering and Land Surveying fundamentals. The author does not guaranty the validity of the content found within this manual. The author of this manual and Santa Rosa Junior College, their heirs and successors will not be held accountable for projects prepared by persons relying on this manual. Knowledge of the material in this manual does not explicitly or implicitly grant a license to perform Civil Engineering and/or Land Surveying. Any questions and/or comments relating to interpretation should be emailed to the author at cest85@aol.com.

© 2007 by Geoffrey J. Coleman. All rights reserved.

No part of this book may be reproduced in any form or by any means without written permission from the publisher, Schroff Development Corp.

The author and publisher of this book have used their best efforts in preparing this book. The efforts include the testing of the tutorials to determine their effectiveness. However, the author and publisher make no warranty of any kind, expressed or implied, with regard to the material contained in this book. The author and publisher shall not be liable in any event for incidental or consequential damages in connection with the use of the material contained herein.

Examination Copies:

Books received as examination copies are for review purposes only and may not be made available for student use. Resale of examination copies is prohibited.

Electronic Files:

Any electronic files associated with this book are licensed to the original user only. These files may not be transferred to any other party.

ISBN: 978-1-58503-354-6

Preface
The Importance of Industry Standardization

As the complexity of construction projects increases and professional collaboration propagates, the need to accurately convey detailed project design information continues to grow. What were once simple line diagrams and verbal agreements between owners, designers and contractors have gradually developed into an intricate network of construction documents. Many of us do not take the time to think about why drawings are represented in the form that they are. Why is there an emphasis on some linework while other linework is less dominant? How does one know how large to annotate objects? How should the sheets in the title block be numbered? We find ourselves turning out drawings and specifications with a particular look, because that's the way our employer has been performing operations for the last 20 years.

With the increasing detail in today's construction projects, we find ourselves developing and implementing our own standards and other organizational procedures. Before developing your own set of standards and procedures, I recommend that you take a look at what's currently available. After all, you could imagine the chaos if every individual preparing documents provided them in a different format.

To increase productivity and bring order to the detail associated with the perplexity of construction documents, a variety of organizations have provided industry standards. The Construction Specifications Institute (CSI) was founded in 1948 by a group of specification writers to improve specification practices in the construction and allied industries by streamlining the exchange of construction data between owners, designers and contractors. In their recent years, the Uniform Drawing System (UDS) was implemented to provide a consistent look for drawings.

Since development of the UDS, the National CAD Standard (NCS), which coordinates CAD publications of multiple organizations has adopted CSI's Uniform Drawing System. The National CAD standard includes a variety of other publications which give suggestions on how manufacturers and designers can organize their drawings and specifications. The publications also provide information on layering and plotting conventions.

Standardizing the way drawings look and function using similar layering and plotting conventions creates a familiarity. This in turn increases productivity and the ability to communicate with other design professionals.

Layering in this manual has been set up to model the layering convention documented by the American Institute of Architects (AIA) endorsed by CSI and NCS. The layering convention makes use of a hierarchy starting with a primary discipline code, a major group, minor group and a status code.

C-WATR-PIPE-EXST
C-WATR-SYMB-NEWW

While the discipline code is a single letter which delineates objects placed on this layer by the design professional, the grouping and status codes contain 4 characters each and work their way from general to specific as modifiers are added to the layer name.

Introduction to Land Desktop 2007

Below you will find examples of Discipline Codes:

Discipline Code	Identification
A	Architectural
C	Civil
E	Electrical
F	Fire Protection
M	Mechanical
P	Plumbing
S	Structural

Below you will find examples of layering under the Civil Discipline:

C-ANNO-DIMS	C-GRND-TOPB	C-ROAD-HPRT
C-ANNO-GRAD	C-NGAS-LINE	C-ROAD-PVMT
C-ANNO-TEXT	C-NGAS-SYMB	C-ROAD-STRP
C-ANNO-REVS	C-PKNG-STRP	C-SDRN-PIPE
C-ANNO-SYMB	C-PKNG-SYMB	C-SDRN-SYMB
C-ANNO-VPRT	C-PNTS-STAK	C-SECT-LINE
C-BLDG-OTLN	C-PNTS-TOPO	C-SSWR-PIPE
C-BLDG-PATT	C-POWR-LINE	C-SSWR-SYMB
C-COMM-OVHD	C-POWR-SYMB	C-SWLK-PATT
C-CONC-PATT	C-PROP-BNDY	C-TREE-DRIP
C-CONC-OTLN	C-PROP-BSLN	C-TREE-TRNK
C-CONT-BNDY	C-PROP-CTLN	C-VGTR-INNR
C-CONT-MAJR	C-PROP-ESMT	C-VGTR-OUTR
C-CONT-MINR	C-PROP-LOTS	C-VMAP-LOCA
C-CONT-SYMB	C-PROP-SYMB	C-WATR-PIPE
C-FENC-BARB	C-PROP-TRAV	C-WATR-SYMB
C-GRND-FLIN	C-ROAD-CTLN	
C-GRND-GBRK	C-ROAD-CURB	
C-GRND-TOEB	C-ROAD-HPLT	

Explanation of Project Folder Assignment Numbers that appear in each exercise:

File naming conventions used in this manual do not stem from the Uniform Drawing System or any other known industry standard. Instead, file names associated with this manual have been set up specifically to aide instructors in grading drawing files in a computer environment. Therefore, the file naming convention implemented in this manual is predicated on instructors assigning unique numbers to students for the purpose of grading files electronically. The implementation of numbers gives each file a unique name, so that multiple files may be recorded to common media for grading while maintaining a record of each students work.

Table of Contents

Lessons **Page**

Reviewing AutoCAD Basics..1.1
Land Desktop Impacts on the AutoCAD Environment....................................2.1
Setting AEC Points... 3.1
Point Groups and the Land Desktop Project Manager.....................................4.1
Importing and Exporting AEC Points..5.1
Rotation, Translation and Datum Adjustment... 6.1
Lines and Curves..7.1
Line Labeling... 8.1
Line and Curve Tables.. 9.1
Point Labeling.. 10.1
Parcel Computations... 11.1
Creating a Surface... 12.1
Faults (Breaklines)...13.1
Creating Contours From Surface Data.. 14.1
Land Desktop Cross Sections...15.1
Earthwork Volumes... 16.1
Alignments.. 17.1
Civil Design Profiles... 18.1
Civil Design Cross Sections...19.1
Civil Design Section Plotting...20.1
Drawing Civil Design Templates..21.1
Defining Civil Design Templates...22.1
Civil Design Control-Templates.. 23.1
Civil Design Control-Alignments.. 24.1
Civil Design Control-Profiles.. 25.1
Civil Design Volume Computations.. 26.1
Civil Design Grading Objects... 27.1
Civil/Survey Tool Pack.. 28.1

Land Desktop Quick Menu Guide

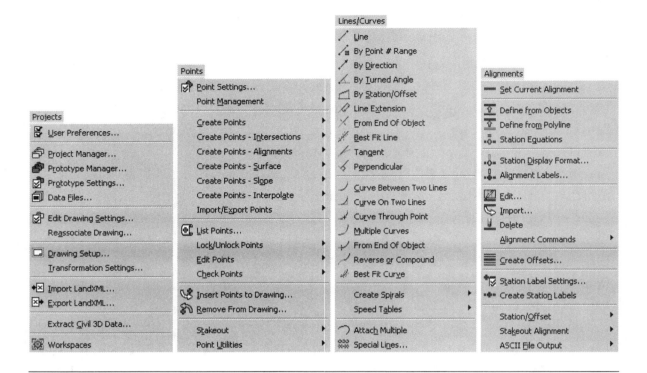
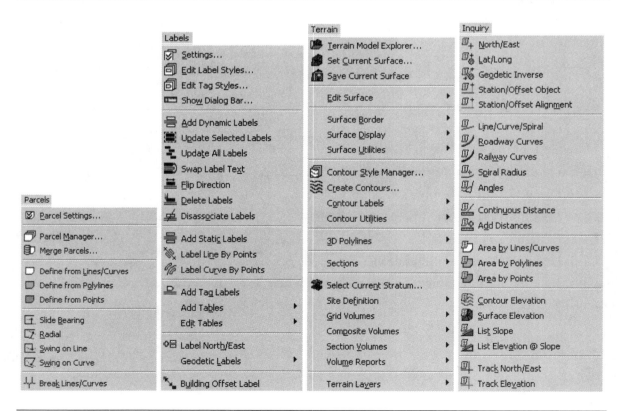

Civil Design Quick Menu Guide

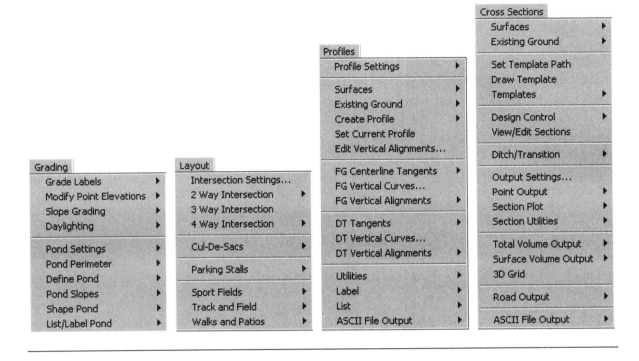

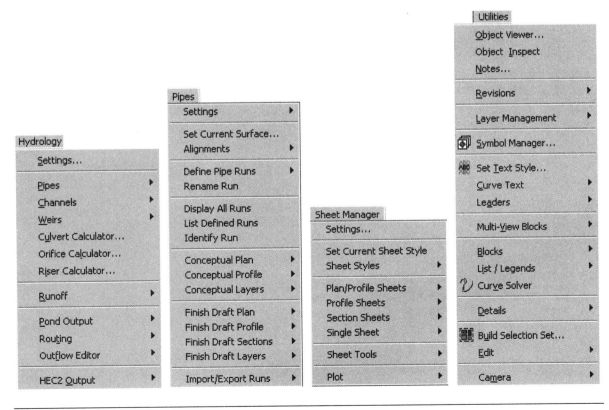

NOTES:

> Assignment #1
>
> ## *Reviewing AutoCAD Basics*

Recommend Assignments Prior to Working this Assignment:

Intermediate AutoCAD

Required Assignments Prior to Working this Assignment:

Beginning AutoCAD

Goals and Objectives

      Technological advances are driving the evolution of the Civil Engineering and Land Surveying fields. Virtually every Engineer and Land Surveyor in urban areas relies on a desktop portable computer (PC) to conduct work. Many local regulatory agencies are requesting submittals on electronic media.

      Autodesk's software AutoCAD, which was at one time perceived as a drawing aide for drafters has since evolved into a network of tools relied upon by design professionals world wide. Land Desktop, previously known as "Softdesk Civil/Survey" (DCA prior to Softdesk), was created to supply Civil Engineers and Land Surveyors with tools for expediting timely tasks related to their discipline. The package that was originally written by an independent party to run within the AutoCAD environment has since been acquired by Autodesk. While AutoCAD has been designed to run as an independent application, Land Desktop will not run without AutoCAD.

      Users who have worked with the Land Desktop software understand the importance of recognizing the difference between AutoCAD and Land Desktop commands. In the next few lessons, you will learn that Land Desktop commands often require different command line syntax than AutoCAD commands. Angles are typically entered with different units, and AutoCAD commands such as "Undo" do not have the same impact on Land Desktop that they do on an ordinary version of AutoCAD.

      If a line is drawn in AutoCAD from the coordinate pair (0,0) to (10,0) to yield a length of 10, then users might conclude that the line length is unitless, and as such the units could perceived as feet, miles or chains for that matter. However, if a user elects to change the drawing unit of measure from "Decimal" to "Architectural" or "Engineering" using the "Format→Units" command, then they will find that AutoCAD interprets the unit of length as "Inches". The previously drawn line now has a magnitude of 10 inches. This is an important concept for Land Desktop users because the base unit for Land Desktop, if imperial units are implemented, is the "US Survey Foot". Land Desktop uses "Feet" as the base unit because the software caters to the Civil Engineering and Land Surveying fields which rely on instrumentation having "Feet" as the base unit. Since there are 12 inches in a foot, objects in architectural drawings, which are typically drawn in AutoCAD, are 12 times larger than objects drawn for the Civil Engineering and Land Surveying disciplines.

      In order to excel using the Land Desktop software, it is relatively important that users have a strong foundation in AutoCAD. Therefore, this lesson has been included to give users the opportunity to review intermediate AutoCAD topics which are critical in the work environment.

All maps contain some form of annotation, so it is important to understand how to set up and manipulate text styles in AutoCAD. In order to be able to set up text styles, users need a general understanding of the components in the text style dialogue box. The **"Text Style"** is merely a name that a user assigns to group of parameters which dictate how the text appears in the drawing environment. The two most important parameters saved with the "Text Style" are the "Font" and "Size" of the text. The **"Font"** dictates the appearance of the text. Below, are examples of different Windows based fonts:

THIS FONT IS KNOWN AS ALGERIAN
THIS FONT IS KNOWN AS GAZE
THIS FONT IS KNOWN AS TIMES NEW ROMAN

The **"Size"** of the text dictates how large the text appears in the drawing. Text heights are assigned in AutoCAD using real dimensions as .08 inches as opposed to most other Windows based applications where the text height is generated by the font pitch.

The "Font" most commonly used by Civil Engineers and Land Surveyors is the "Simplex". **"Simplex"** is used, because it can be replicated manually with very little effort and it doesn't consume very much memory in an AutoCAD file.

The most popular text size used by Civil Engineers and Land Surveyors is .10 inches tall on the paper plot. This is a popular size because it can still be read after generating half scale reproductions.

You have most likely learned in your elementary AutoCAD class that **drawing scale** throws an interesting twist on the appearance of text printed on the paper plot. Since the objects are shrunk down by your drawing scale to fit them on the paper plot, it is necessary to enlarge your text by the same scale that your objects were reduced by to achieve the desired .10 inch text height on the paper plot. This means that a drawing plotted out 10x smaller than the actual drawn model space objects needs to have the text enlarged by a factor of 10 to maintain .10 inches on the paper plot. In the previous example, in order to maintain .10 inches on the paper plot, the text size in the model environment would need to be 1.0 inches tall.

Although AutoCAD permits users to name styles just about whatever they want, its generally a good idea to implement names that are going to mean something to the next person that picks up your drawing. As an example, AutoCAD would permit a user to set up a text style with the name "Ryan" making use of the "Simplex" font and a height of 1.0 inches. As mentioned above, this text style would appear perfect on a 10 scale (1in = 10ft) drawing prepared by an Engineering or Surveying Firm. However, the name "Ryan" does not tell us anything about the text style parameters.

It just so happens that someone has come up with a naming convention that gives users information about the text style that they are implementing. This naming convention is known as the "Leroy". The **"Leroy"** text style makes use of the simplex font and is coded in a fashion which tells the users how large the text should appear on the paper plot. Common examples of the Leroy text style are the L60, L80, L100, and

L120. The L80 text style should be set up to yield a text height of .08 inches on the paper plot. The L100 text style should be set up to yield a .10 inch height on the paper plot. This means that a user must manipulate the height of the text in the drawing by the drawing scale so that the text plots on paper to the correct height.

Unfortunately, the Leroy text style does not do a great job of catering to electronic drawings with multiple scales. An L100 at a 20 scale would require a model space height of 2.0 inches, while an L100 at a 10 scale would only require a model space height of 1.0 inches, yet both text styles are named L100. In the case where multiple scales are implemented in a single drawing file, consider a modifier which reflects the scale of the drawing. The name L100-020X indicates that the particular text style is a Leroy (Simplex Font) the paper text height is .10 inches, and the electronic text style is set up for a 20 scale drawing with an absolute height of 2 inches. Several examples of the amended Leroy naming convention are listed below for reference.

Style	Ht.	Style	Ht.	Style	Ht.
L080-001X	.08	L080-010X	.80	L080-020X	1.6
L100-001X	.10	L100-010X	1.0	L100-020X	2.0
L120-001X	.12	L120-010X	1.2	L120-020X	2.4

For this lesson, we will begin our first project. While the remainder of this manual focuses on the Civil Engineering and Land Surveying trades, this lesson is an excerpt from an Architectural floor plan. This lesson will allow you to observe the relationship between Architectural and Civil/Survey drawings, exercise your AutoCAD skills and give you the opportunity to become more familiar with your workstation. In particular, this lesson will target proper text and dimension style setup and basic floating viewport configuration using layout tabs.

Exercise Instructions

- Logon to your workstation and begin a session of Land Desktop. If Land Desktop has been configured on your workstation to display the "Start Up" dialogue box, then cancel this feature so that the AutoCAD model space environment is displayed. If launching Land Desktop brings you directly into the AutoCAD model space environment, then Land Desktop has been configured to begin without the startup dialogue. [Users that have the ability to customize their profiles can turn this feature off permanently through the menu "Tools→Options" or by typing "OP" at the command line. In order to toggle this option off permanently, select the "System" tab, and specify "Do not Show a Startup Dialogue" for the drop down list under the "Startup" section on the Right side of the dialogue box.] If the AutoCAD Map "Task Pane" is shown, then close this dialogue box as well. [If users have the ability to customize their profiles, then this feature can be turned off permanently through the menu "Map→Options" and toggling off the checkbox adjacent to "Show Task Pane on startup".]

- Begin a new drawing by selecting "File→New" and observe the dialogue box which appears. The "Project Path:" reveals the location where projects will be created and stored. The "Project Name:" reveals the name of the project which newly created drawings will be associated with. The "Drawing Path:" reveals the directory structure which a user would need to follow to locate drawings associated with the project.

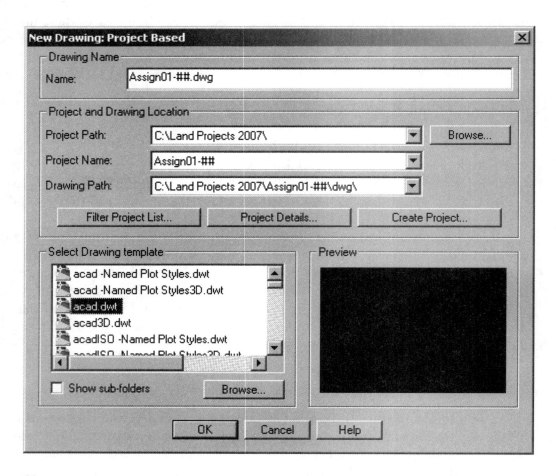

- Choose to create a new project by picking the button labeled "Create Project". Select the "Prototype:" "Default (Feet)" from the drop down list. Give the project the "Name:" "Assign01-##" where the "##" reflects the number which was assigned to you on the first day of class. If the number assigned to you is a single digit (for example 4 as opposed to 14), then enter a zero preceding the single digit (such as 04). Write "Reviewing AutoCAD Basics" in the "Description:" and "Keywords:" areas. This will serve as a project summary which may later be used to find or filter projects using the Project Manager. Refer to the graphic on the following page for reference.

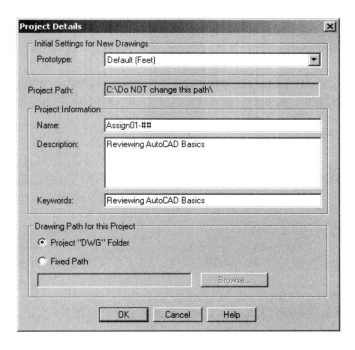

- Select "OK" to bring you back to the "New Drawing" dialogue box and type the "Name:" "Assign01-##" where the "##" reflects the number which was assigned to you on the first day of class. The "Project Path:" may vary based on the software installation and intended location of project data. The paths shown in the graphic above reflect the paths users will encounter with the default Land Desktop installation.

- Select the "acad.dwt" template and then the "OK" button to generate a new drawing in the "DWG" folder of the current project. After selecting the "OK" button, you may be prompted to save changes to the previous drawing session. Since there were no objects in the previous drawing session worth saving, choose not to save changes.

- The relationship between Land Desktop and AutoCAD is an extremely important concept. We will see in future lessons that Land Desktop alters Text and Dimension Style geometry. In the current assignment, it is important that users thoroughly understand how to set up text and dimension styles without the aid of Land Desktop. In order to accomplish this, we will cancel the Land Desktop setup routine that sets up text styles for us and generate them manually. These manually created text styles will then be incorporated into our dimension styles.

- You will be prompted with the "Load Settings" dialogue box. **CANCEL** this dialogue box using the "**X**" in the upper right hand corner to minimize the impact that Land Desktop has on the AutoCAD drawing environment.

- You will be prompted with the "Create Point Database" dialogue box. Accept the default settings by choosing "OK".

- Before we begin drawing, it is important that good layer management be implemented. While mocking AIA's layering convention, set up 7 layers in the following manner.

Layer Name	Color	Linetype
A-ANNO-DIMS-010X	Green	Continuous
A-ANNO-DIMS-100X	Red	Continuous
A-ANNO-TEXT	White	Continuous
A-ANNO-VPRT	White	Continuous
A-CNTR-LINE	Cyan	Center
A-KCHN-SYMB	Yellow	Continuous
A-WALL	Magenta	Continuous

- Set up text styles as discussed earlier in this manual using the Leroy 100 at a unit scale (Named L100-001X), 10scale (Named L100-010X) and 100scale named (L100-100X). The text style dialogue box may be invoked using the menu "Format→Text Style" or by typing "ST" at the command prompt. Keep in mind that by definition a Leroy 100 text style is a simplex font with a plotted height of .10 inches tall. This means that the height of a Leroy 100 text style at a unit scale is .10 inches tall. A 20 scale drawing has a Leroy 100 text style with a height of 20 * .10 inches = 2 inches. The text style is scaled up so that when the drawing is scaled down by a factor of 20 for the paper plot, the text height on the paper will be .10 inches tall.

- Set up dimension styles using the Leroy 100 at 10scale (Named Dims-L100-010x) and 100scale (Named Dims-L100-100x). The Dimension Style Manager may be invoked using the menu "Format→Dimension Style" or by typing "D" at the command prompt. Refer to the section at the end of this chapter titled "Ballpark values for setting up Dimension Geometry" to aide in setting up these dimension styles. Keep in mind that the dimension style shown in the graphics, at the end of this chapter, has been set up for a model space environment intended to be plotted at the imperial scale 1in=20ft (20 scale).

- Make sure that the "Model" layout is active and begin drafting the building footprint using the illustrations on the following pages. For this assignment, the building footprint and counter should be drawn on the layer "A-wall". The appliance should be constructed on the layer "A-kchn-symb" (kitchen symbols). Construction centerlines laying out the gas range should be drawn on the layer "A-cntr-line" (Centerline).

- After your building footprint, gas range and counter are drawn, dimension the gas range using the 10scale dimension style and the building footprint using the 100scale dimension style as shown in the following illustrations.

Reviewing AutoCAD Basics

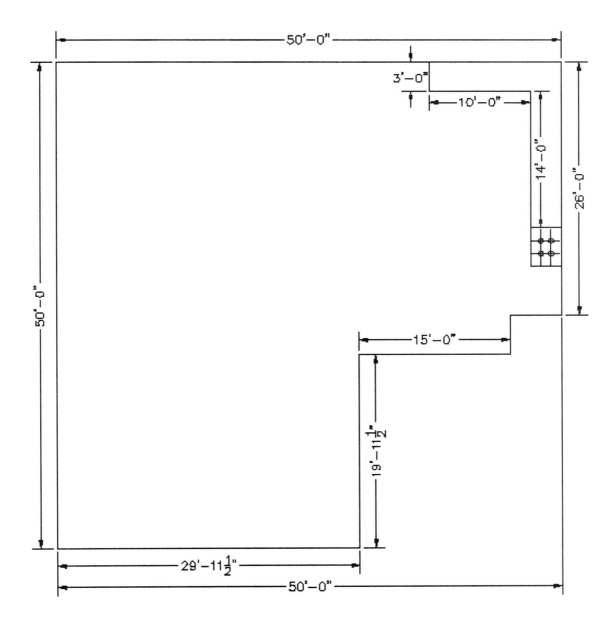

1.7

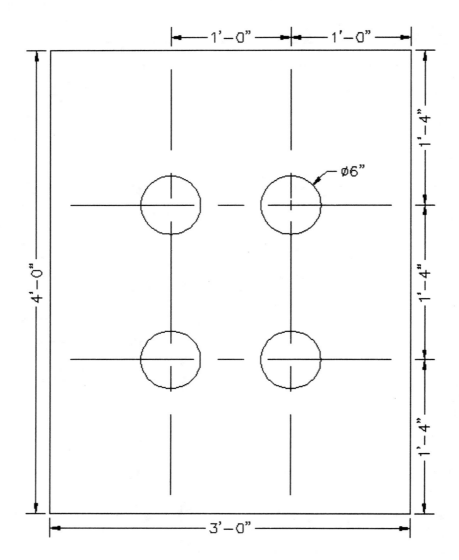

- The illustration on page 1.10 is a non-scale plot of two floating (Paperspace) viewports. Creating these viewports will be our next task. Set tilemode to 0, moving to the paperspace environment, by selecting the layout tab labeled "Layout1". Right select on this Layout and choose to rename it to "Architectural Plan". The administrator who installed the Land Desktop software on the PC which you are working may have configured it to automatically generate paperspace viewports when layout space is invoked. If a viewport has been created automatically, then erase it using the AutoCAD "Erase" command. While on the "A-ANNO-VPRT" layer create a floating viewport by typing "Mview" at the command line. Start your viewport at the origin (0,0) and choose the coordinate pair (11,8.5) for the other corner. Create a second viewport, the same size as the first viewport, 1 inch above the previous viewport. Set the scale of the lower viewport by picking the "Paper" button in the task bar, so that the button changes to read "Model," and zooming the viewport to a scale of 1/100xp. Other popular methods for setting the viewport scale in AutoCAD are through the implementation of the "MVSetup" command and through the "Properties" dialogue box. Set the scale of the upper viewport by zooming to a scale of 1/10xp.

- Use a viewport freeze to hide the layers that we do not wish to show up in their respective viewports. As an example, the 10 scale dimensions should not show up in the 100scale viewport.

- Your centerlines might look like continuous lines while in the paperspace (Layout Tabs) environment. Since this is the environment that we would be plotting from, if we were to plot this assignment, we want the centerlines to look correct in this view. If we set the AutoCAD linetype scale (LTSCALE) to 1 and the AutoCAD system variable "PSLTSCALE" to 1, then the linetype scale will take on the viewport scale for each viewport. This means that our lines will appear correct in any and all viewports regardless of the scale, provided the properties of individual entities have not been set to values other than 1. For this reason, it is important not to change the linetype scale for individual entities using the "Change", "Ddmodify", "Properties", "Entity Display" or similar commands. Set the variables "LTSCALE" and "PSLTSCALE" to 1.

- While in paperspace, use the Multiline Text Editor to type your name between the two viewports using the L100 text style set up at the unit scale.

- Save and exit your drawing. You have successfully completed your first assignment.

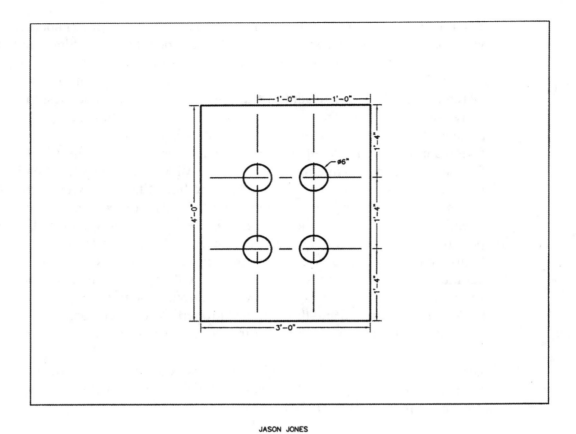

Reference Figure: Assignment 1

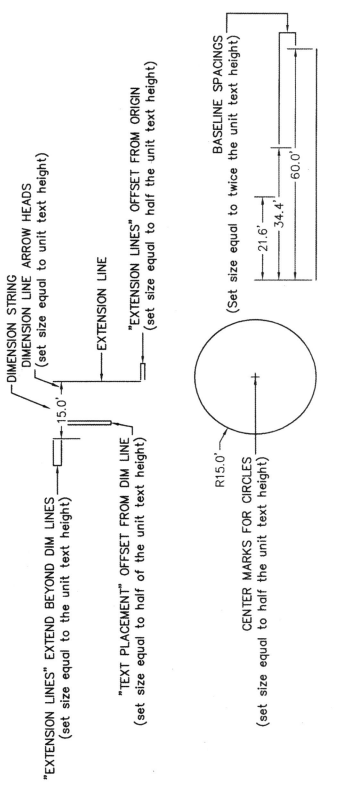

THE DIMENSION STRING TEXT HEIGHT IS GOVERNED BY THE TEXT STYLE PARAMETERS OF THE "TEXT STYLE" SPECIFIED IN THE "TEXT" AREA OF THE "DIMENSION STYLE MANAGER" UNLESS THE ASSIGNED "TEXT STYLE" HAS A HEIGHT DEFINED AS 0 IN THE "TEXT STYLE" DIALIGUE BOX. IF THE "TEXT STYLE" HAS AN ASSIGNED HEIGHT EQUAL TO 0, THEN THE DIMENSION STRING TEXT HEIGHT WILL TAKE ON THE VALUE SPECIFIED AS "TEXT HEIGHT" IN THE "TEXT" AREA OF THE "DIMENSION STYLE MANAGER".

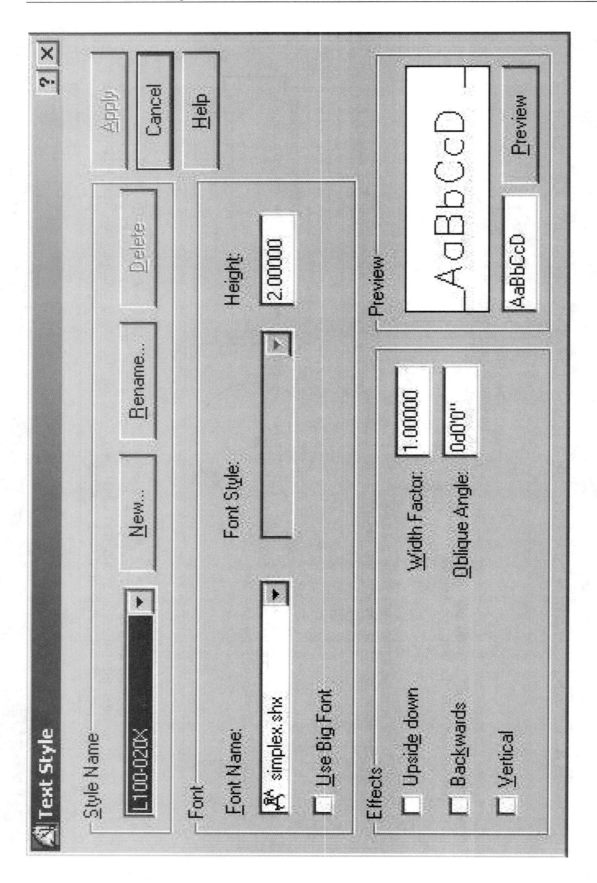

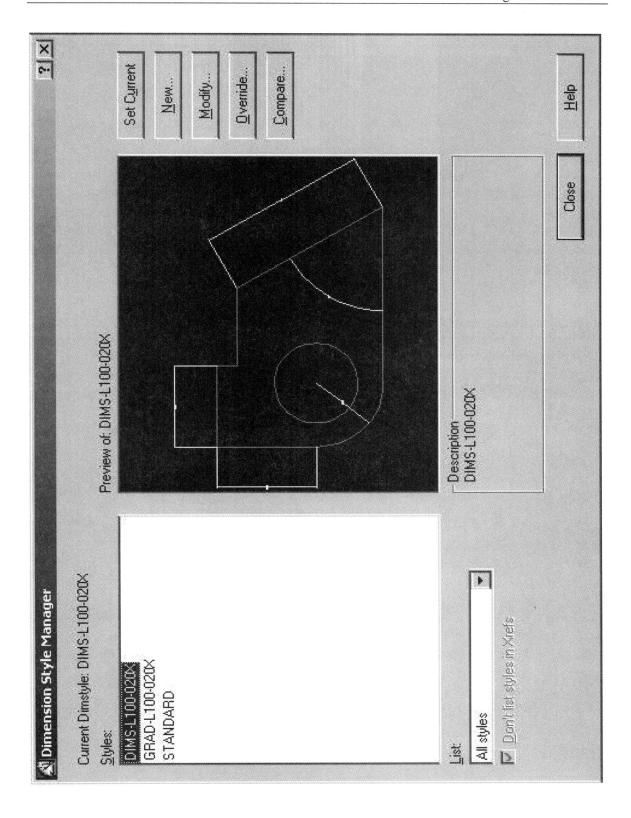

Introduction to Land Desktop 2007

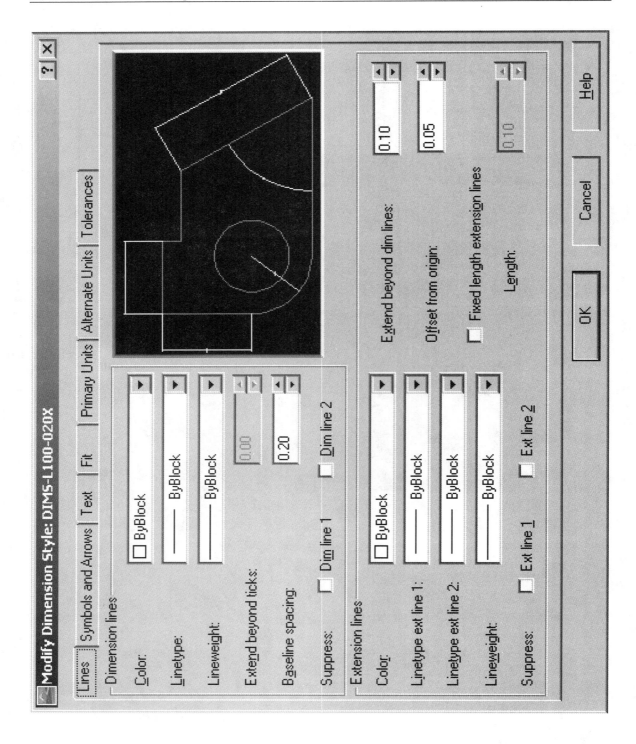

1.14

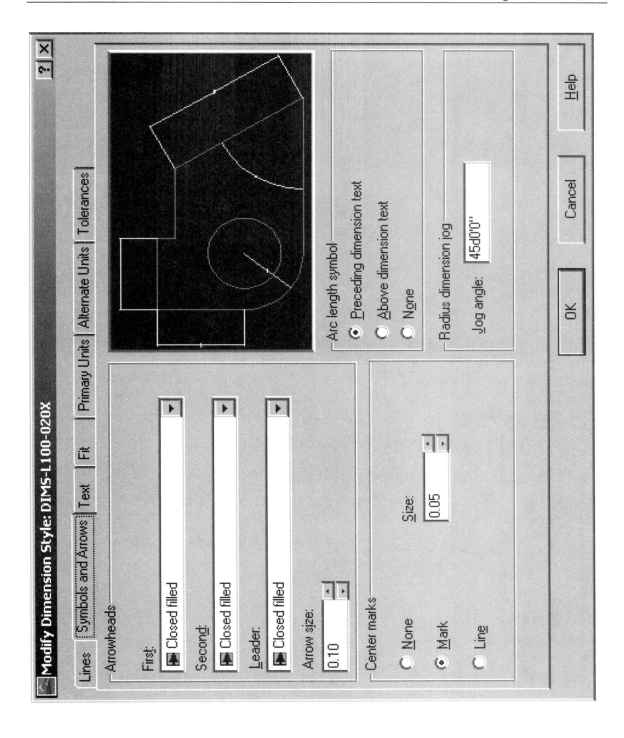

Introduction to Land Desktop 2007

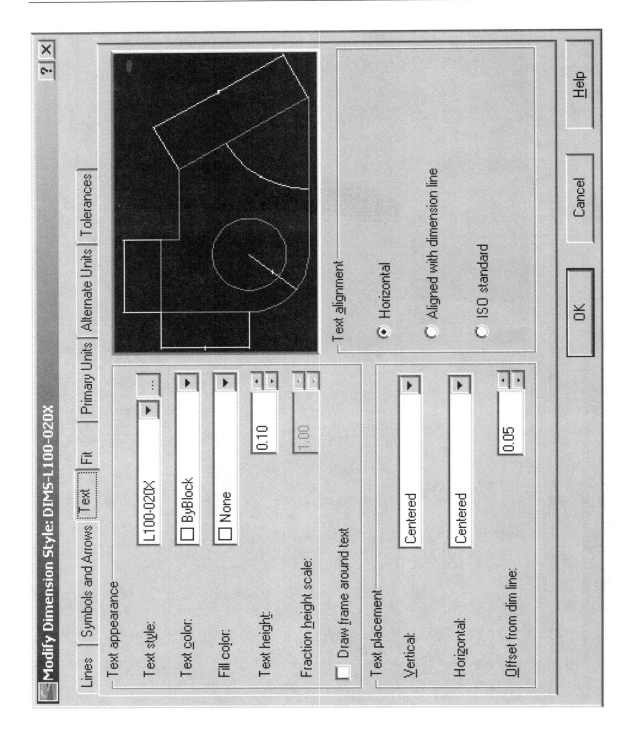

1.16

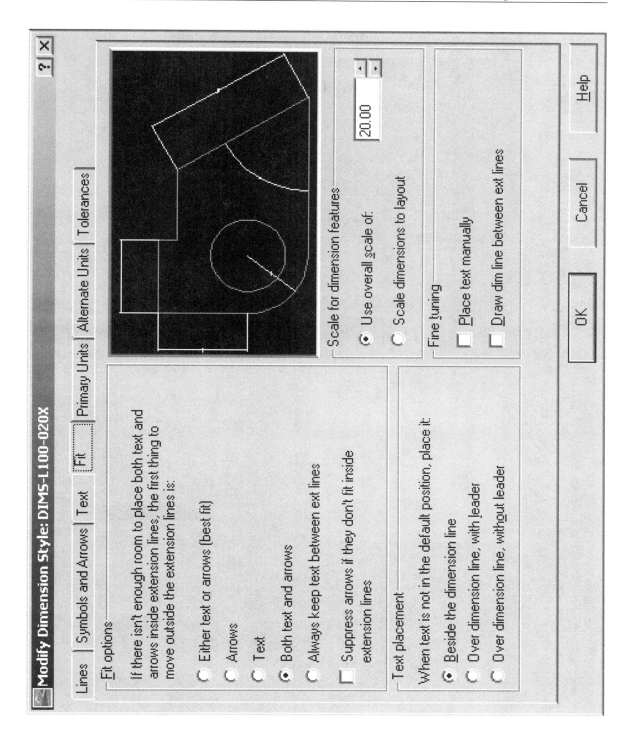

Introduction to Land Desktop 2007

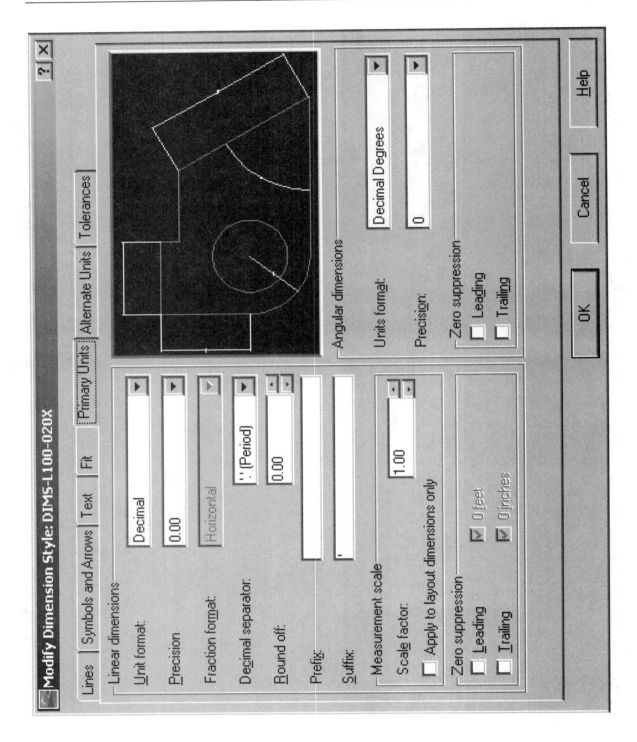

1.18

> Assignment #2
> # *Land Desktop Impacts on the AutoCAD Environment*

Recommended Assignments Prior to Working this Assignment:

Assignment 1

Required Assignments Prior to Working this Assignment:

Assignment 1

Goals and Objectives

      The Land Desktop software requires that drawings be saved before many of the Land Desktop commands can be invoked. This is because some of the Land Desktop commands generate external (separate from the AutoCAD drawing) data files composed of information which relate to drawings. In an effort to keep these data files together, Land Desktop requires that users provide a name, so that this information can be packaged and stored in folders called projects.

      While users generally do not access the external data files directly, the software reads and writes to them as commands are invoked. Therefore, organization is of the essence and users should not attempt to move or erase files using utilities such as "Windows Explorer". A "Project Manager" has been included with the Land Desktop Software which we will discuss in future assignments.

      Although it is important that the software read and write to a single location (project) for each AutoCAD drawing, some of the external data files may be accessed by multiple drawings. For this reason, a single project may be composed of several drawing files, but the drawing files can only reference a single project.

      Separating drawing components into multiple AutoCAD drawing files reduces file size which increases software performance. The addition of external data files increases productivity because it allows users to access and manipulate data which is common to all drawings located within a project.

      As Land Desktop users, you will find that it is good practice to insert a drawing into an environment that has all ready been set up by Land Desktop as opposed to running the drawing setup on an AutoCAD drawing that hasn't been manipulated by Land Desktop. In order to make some sense of the previous statement, consider the following example.

      As a survey technician, you have been assigned the task of providing construction calculations for your survey crew so that they may stake the corners of a proposed residential structure. Suppose that the electronic file will be coming from a firm that does not use the Land Desktop software. You seem to remember learning about the impact of Land Desktop on AutoCAD files from a class that you took a year ago and decide that you want to see some of the impacts for yourself. You open the file up in a stand-alone copy of AutoCAD 2007 and find a pretty simple drawing with a bit of linework, dimensions and text. You find that the drawing contains 2 text styles and a dimension style. One of the text styles reads "Standard" and has a text height of 4 units. The other text style reads "CHRISTXT" and has a height of 8 units. The dimension style is also named "Standard" and makes use of the text style "Standard." After making some notes to yourself, you decide to open the file up in Land Desktop as opposed to setting up a new drawing in Land Desktop and inserting the provided drawing into the environment that has all ready been set up. After opening the drawing in Land Desktop, you run through the drawing setup and choose a horizontal scale of 10 and a text style of L80. When you go back and check the text styles, you don't find anything wrong with the new text styles that Land Desktop has set up, but you find that the existing text styles have been modified. The "Standard" text style now has a height of 1 unit and the "CHRISTXT" text style has been assigned a height of 2 units. You also find that overrides have been applied to the "Standard" dimension style. If you would have created a new drawing, set up the new drawing in Land Desktop and then inserted the architectural drawing into the newly created drawing environment, you would not have had this problem. The text and dimension styles of the inserted drawing would have maintained their correct heights of 4 and 8 units tall.

Land Desktop includes a setup wizard which adjusts many of AutoCAD's default settings. This assignment has been crafted to allow users to observe the impacts Land Desktop has on the AutoCAD environment. Users will first launch a session of Land Desktop, execute a series of AutoCAD commands, and record the command default values without using the Land Desktop setup wizard. Users will then create a new drawing with the aide of the setup wizard, execute the same commands and compare the differences in values. In particular, we will learn how setting up a project impacts AutoCAD text styles, dimension styles, units, limits, precision and the global line type scale.

Exercise Instructions

- Logon to your workstation and begin a session of Land Desktop. If Land Desktop has been configured on your workstation to display the "Start Up" dialogue box, then cancel this feature so that the AutoCAD model space environment is displayed. If launching Land Desktop brings you directly into the AutoCAD model space environment, then Land Desktop has been configured to begin without the startup dialogue. If the AutoCAD Map "Task Pane" is shown, then close this dialogue box as well. Users that have the ability to customize their profiles can turn these features off permanently by following the procedure described in Assignment #1.

- Begin a new drawing by selecting "File→New". Choose to create a new project by picking the button labeled "Create Project". Select the "Prototype:" "Default (Feet)" from the drop down list. Give the project the "Name:" "Assign02-##" where the "##" reflects the number which was assigned to you on the first day of class. If the number assigned to you is a single digit (for example 4 as opposed to 14), then enter a zero preceding the single digit (such as 04). Write "Land Desktop Setup" in the "Description:"

and "Keywords:" areas. This will serve as a project summary which may later be used to find or filter projects using the Project Manager.

- Select "OK" to bring you back to the "New Drawing" dialogue box and type the "Name:" "Assign02-##" where the "##" reflects the number which was assigned to you on the first day of class. Make sure that the correct project is specified in the "Project Name:" area.

- Select the "Acad.dwt" template and then "OK" to generate a new drawing in the "DWG" folder of the current project. After selecting the "OK" button, you may be prompted to save changes to the previous drawing session. Since there were no objects in the previous drawing session worth saving, choose not to save changes.

- You will be prompted with the "Load Settings" dialogue box. **CANCEL** this dialogue box using the "**X**" in the upper right hand corner to minimize the impact that Land Desktop has on the AutoCAD drawing environment.

- You will be prompted with the "Create Point Database" dialogue box. Accept the default settings by choosing "OK".

1. What are the LIMITS of the current drawing? (0,0) and _____

2. What is the Global Linetype Scale (LTS) of the current drawing _____

3. Select "Format→Text Style" to invoke the "Text Style" dialogue box. If the "Format" menu is not available, then change "Workspaces" by selecting the Workspace "Land Desktop Complete" from the drop down list in the "Workspaces" toolbar. If the "Workspaces" toolbar is not displayed on the screen, then it can be invoked using the command "Projects→Workspaces". ["Workspaces" is a new feature that was introduced in the AutoCAD 2006 suite of products and has taken the place of the "Menu Palette Manager" used in previous versions of Land Desktop.] What text styles, if any, are available for this new drawing session? _____ Close the style dialogue box.

4. Select "Format→Dimension Style", select the "Standard" style and then "Set Current→OK" to eliminate overrides that may have been created by the start of the Land Desktop setup. Select the "Modify" option. Is there anything displayed in the "Suffix" area on the "Primary Units" tab? _____

5. Go to the "Text" tab. What is the current text style associated with the default dimension style? _____

6. Select the "Fit" tab. What does the "Use overall scale of" read? _____ Exit the dimensions style manager.

7. Select "Format→Units". What is the "Precision" set to? _____

- Begin a new drawing by selecting "File→New." Accept the default project which will be the current project "Assign02-##" and type the "Name:" "Assign02-##" where the "##" reflects the number which was assigned to you on the first day of class. If the number assigned to you is a single digit (for example 4 as opposed to 14), then enter a zero preceding the single digit (such as 04). Make sure that the "Project Name:" displays the correct project.

- Select the "Acad.dwt" template and then "OK" to generate a new drawing in the "DWG" folder of the current project. Choose to overwrite the existing file when prompted. Since there were no objects in the previous drawing session, choose not to save changes.

- You will be prompted with the "Load Settings" dialogue box which displays a list of drawing setups to choose from. These setups are saved back to the path determined by the Network Administrator who installed the software. The default path is to the local machine. We will set up our parameters manually, so select "Next" to set up your project with the following parameters:

    Units Area
        Linear Units = Feet
        Angle Units = Degrees
        Angle Display Style = Bearings
        Display Precision Linear = 0
        Display Precision Elevation = 2
        Display Precision Coordinate = 5
        Display Precision Angular = 4
        "Next"

    Scale Area
        Horizontal = 100 [This option is displayed as 1"=100']
        Vertical = 1 [This option is displayed as 1"=1']
        Paper Size = 8 x 11(A)
        "Next"

    Zone Area
        "Next"

    Orientation Area
        "Next"

    Text Style Area
        Leroy.stp
        L100
        "Finish"

- A screen will display providing the user with a summary of the settings chosen. Review the settings and select "OK."

8. What are the LIMITS of the current drawing? (0,0) and _____
9. What is the Global Linetype Scale (LTS) of the current drawing _____
10. Select "Format→Text Style". What is the current text style? _____

11. List three text styles, available in this new drawing session? _____ Close the style dialogue box by selecting the "Cancel option".

12. Select "Format→Dimension Style," select the "Style Overrides" style then the "Modify" option. Is there anything displayed in the "Suffix" area on the "Primary Units" tab? If so, what is displayed? _____

13. Go to the "Text" tab. What is the current text style associated with the default dimension style? _____

14. Select the "Fit" tab. What does the "Use overall scale of" read? _____

- Exit the dimensions style manager. Select "Insert→Block→Browse" to locate and insert assign01-##.dwg in to modelspace at the coordinate pair (0,0) with a rotation angle of 0 and a scale factor of 1. Zoom Extents and explode the block one time. If you elect to explode the block upon insertion using the check box in the "Insert" dialogue box, then exploding it a second time is not required or recommended.

15. Select "Format→Text Style" and look at the text height associated with your text style L100-100X. Explain how its height compares to the height of the L100 text style? _____
_____
_____

16. Select "Dimension→Leader" and pick two points on the screen to specify a leader orientation. Press the "Enter" key three times to toggle through the command line defaults and get to the Multiline Text Editor. Type in your name and select "OK." Using the "Tools→Inquiry→List" command the height of the text is _____

17. Select "Format→Units". What is the "Precision" set to? _____

- Save and exit your drawing. You have successfully completed this assignment.

NOTES:

> Assignment #3
>
> ## *Setting AEC Points*

Recommended Assignments Prior to Working this Assignment:

Assignments 1-2

Required Assignments Prior to Working this Assignment:

None

Goals and Objectives

      The first step in generating maps for the Civil Engineering and Land Surveying professions often begins with researching a combination of title information, the general plan, zoning and public records. In having an understanding of the area requiring a survey, and the regulations specific to the area being mapped, surveyors have a better feeling for locating significant site features.

      A deed to the subject property, which is a form of title information, might reveal that iron pipes demark the corners of the parcel being surveyed. Locating and mapping these pipes often aides Land Surveyors in determining an accurate boundary. A sanitary sewer manhole in the street adjacent to the subject property, which may be have been shown on an existing set of public improvement drawings, would reveal the location of the public sanitation facilities. Studying these aspects of a project in advance gives Engineers and Surveyors a feeling for the types of site features to search for and map.

      Site survey information can be located through the implementation of a Theodolite and Chaining, using a Global Positioning System, through Photogrammetry, or using a Total Station. Actually, these are just a few of the many conventional means of locating site information.

      The Global Positioning System and Total Station allow Surveyors to store data relating to site features in a computerized data collection device so that this information can be uploaded to a personal computer for analysis and reduction. Although many types of information may be stored in these microcomputers referred to as data collectors, some of the more common pieces of information stored are the horizontal and vertical location of site features, as well as a description of the feature being mapped.

      When these data components are uploaded to a personal computer using Land Desktop, they are read into an external database in the form of points. The horizontal placement is dictated by its "Y" (Northing) and "X" (Easting) components. The vertical placement in the database is dictated by the elevation stored in the data collection device.

      As was discussed in previous chapters, there are advantages to storing point data in an external database. Not only does this help reduce the AutoCAD file size thereby increasing system performance, but it also gives multiple users who are working on the same project the opportunity to access data without the other having to exit a drawing. Of course, in order to take advantage of this powerful feature, the database must be set to allow multiple users. This setting can be altered through the menu "Points→Point Management→Point Database Setup".

This lesson has been included to give users experience creating points manually in the database/drawing environments. Land Desktop users should adjust the point settings using the menu "Points→Point Settings" prior creating points in the database/drawing environments. Understand that these settings only impact points which are going to be set, or re-inserted into Land Desktop. Points which are all ready in Land Desktop may have their settings adjusted using the "Points→Edit Points→Display Properties" command. The following dialogue box displays the "Create" tab in the "Point Settings" dialogue box.

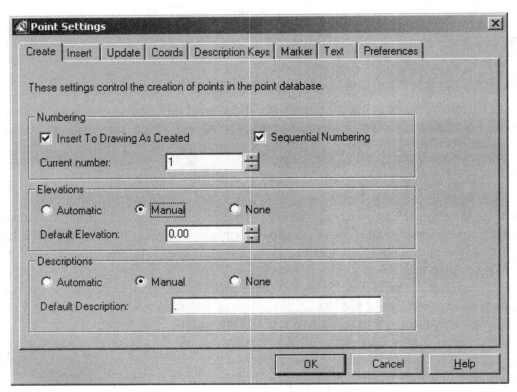

Points which are generated in the database do not necessarily need to be inserted into the current drawing environment. The AutoCAD drawing environment gives users the ability to generate maps and observe the data. Unlike older generations of the software, where points could be queried in the drawing or database, the Land Desktop commands presently only rely on the information in the external database. As such, drawings are a tool for the user and it is not necessary that the data be located in the drawing for the Land Desktop software to perform its computations.

Points which are inserted into the AutoCAD drawing environment can include any combination of the point marker which is located at the Northing and Easting stored in the external database and three data components, sometimes referred to as attributes (because they were block attributes in earlier generations of the software). Any combination of the data components may be displayed. Among them are the point number, the point elevation and the point description.

In the dialogue box shown above, if the "Insert To Drawing As Created" check box is turned on, then points will be inserted into the drawing when generated in the database. This includes points being imported into projects.

Points which are generated in the database are assigned a unique point number. The "Current number:" represents the number which will be assigned to the next point created in the database.

If the "Sequential Numbering" check box is toggled on (Checked), then the software will automatically increment the point number for the next point created. If the check box is turned off, then the software will prompt a user for the desired point number. Users more commonly leave this setting turned on.

Elevations can be set three different ways. If set to "Automatic", then the software will display the elevation typed in the "Default Elevation:" area every time points are created, until the settings are changed. This is handy when users desire to create multiple points with the same elevations. If "Manual" is chosen, then the software will prompt users for elevations as points are generated. If "None" is selected, then the software simply will not log an elevation for points created after this setting is changed.

The default raw description length is set up by users when a project is first created. The default is 32 characters and is generally a lot more than users want to implement. Experience will demonstrate that long descriptions generate more work for users and they have a tendency to clutter maps. The description settings work similar to elevation settings. If set to "Automatic", then the software will display the description typed in the "Default Description:" area every time points are created, until the settings are changed. This is handy when users desire to create multiple points with the same descriptions. If "Manual" is chosen, then the software will prompt users for descriptions as points are generated. If "None" is selected, then the software simply will not log a description for points created after this setting is changed.

The "Elevations" and "Descriptions" areas do not apply for points being imported into projects. Instead, the elevation and description data components are controlled by the parameters in the import file. Chapter 5 discusses importing and exporting project points in more detail.

The following dialogue box displays the "Insert" tab in the "Point Settings" dialogue box.

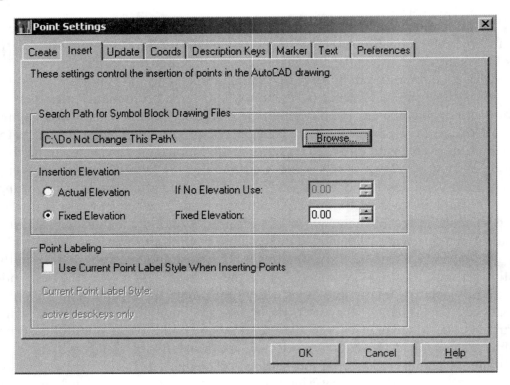

The "Search Path for Symbol Block Drawing Files" shows the path for symbol blocks which are used in conjunction with "Point Labeling" and "Description Keys" discussed in a later assignment.

Points can be inserted in the AutoCAD drawing environment at their actual elevation or at a user specified elevation. Although we are mapping and designing elements in a 3-dimensional world, conventional mapping is still performed 2-dimensionally. Therefore, users will want to leave the "Insert Elevation" setting to yield a "Fixed Elevation" of "0.00". This way, when lines are inquired after drawing them in AutoCAD from one node to another, they yield 2-dimensional distances.

The "Point Labeling" and "Description Keys" settings will be addressed in a later assignment when these topics are covered.

The following dialogue box displays the "Update" tab in the "Point Settings" dialogue box.

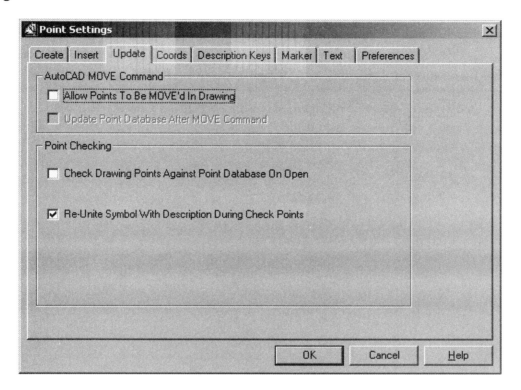

Points which are moved using the "Points→Edit Points→Move" command are automatically moved in the drawing and database. If the "Allow Points To Be MOVE'd In Drawing" is toggled on, then Points which are in the AutoCAD drawing environment can also be moved using AutoCAD commands. If the check box is NOT toggled on, then the data components can be moved, but the marker will remain fixed. The more common setting is to leave this setting toggled off to prevent users from accidentally moving points in the drawing environment. While this setting can be toggled back on to allow points to be moved using AutoCAD commands, there are circumstances when the software will remember information relating to the point and not allow the point to be moved using AutoCAD commands, even if the check box is toggled to do so. If this occurs, simply toggle on the setting, and perform an AutoCAD "Regeneration all" by typing "REA" at the command line. Toggle the setting off and perform an AutoCAD "Regeneration all" (REA), and then toggle the setting back on again. This will cache the point settings and allow users to move points using AutoCAD commands.

Points which are moved in the AutoCAD drawing using AutoCAD commands will not be moved in the external database unless the check box "Update Point Database After MOVE Command" is toggled on. Another form of updating the project point database, referred to as "Check Points", will be discussed in a later assignment.

If the "Check Drawing Points Against Point Database On Open" check box is toggled on, then Land Desktop will analyze differences between points in the drawing and external database when drawings are opened. If there are differences, then the user will be prompted as such.

The "Re-Unite Symbol With Description During Check Points" allows point labels and blocks inserted with "Description Keys" to be relocated to their correct position as points are moved in the AutoCAD drawing.

The following dialogue box displays the "Coords" (Coordinates) tab in the "Point Settings" dialogue box.

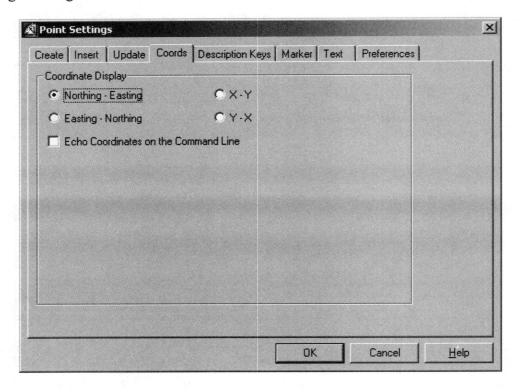

The "Coordinate Display" allows users to choose how they prefer to observe data when inputting point related information and when performing Land Desktop inquires. The more common setting is to display "Northing-Easting".

"Echo Coordinates on the Command Line" will allow users to view point data at the AutoCAD "Command Line" as it is keyed in and as inquires are made.

The "Description Keys" tab will be addressed in a future assignment, so we will skip over these settings for the time being.

The following dialogue box displays the "Marker" tab in the "Point Settings" dialogue box.

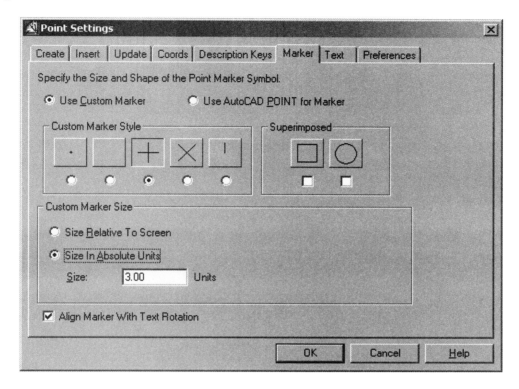

In previous releases of the software, point markers were restricted to AutoCAD points. The Land Desktop software now allows users to implement custom markers. An advantage to using the custom marker is that we can reserve AutoCAD points for other uses. One of the big concerns users first had with custom point markers was whether or not the AutoCAD object snap "Node" works with custom point markers. Users should not be alarmed, because the "Node" object snap still functions with custom point markers.

The most commonly implemented marker setting is the "+" marker. Maps tend to read well when the marker size has a height equal to your model space point data component text. In order to actually achieve this, users must set the marker height to ½ the model space point data component text height. If points are going to be read into the drawing environment using an L060_100X text style (Model space text height = 6.0), then the point marker height should be set to 3.0 units so that the resulting marker and point data component text are the same height.

Many map creators tend to implement a slight rotation on spot elevations (Points) so that the data may be easily depicted by map readers. Users will want to toggle on the option to "Align Marker With Text Rotation" so that the marker aligns with the text.

The following dialogue box displays the "Text" tab in the "Point Settings" dialogue box.

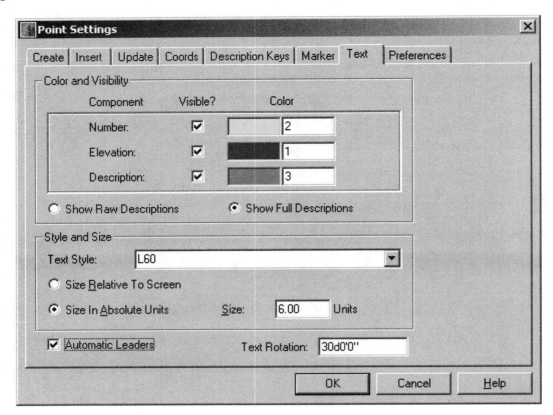

Each of the "Number", "Elevation" and "Description" data components can be turned on or off by selecting the "Visible" check box adjacent to each of the data components. Users can choose the color of each component by typing in a number representing the color of choice or by picking the color adjacent to the component and selecting a color from the color chart.

Users have the ability to choose "Full" or "Raw" descriptions. These features apply to description mapping which will be discussed in a future assignment.

The text style used should be set up for the model space environment and should have an absolute text size which will accommodate the drawing scale. The "Text Style" L060-100X has a model space height of 6.0. "Text Styles" displayed in this area are generated in the AutoCAD "Text Style" dialogue box.

When a point's data components are moved away from the marker, often times to make them more legible due to overlapping data components of adjacent points, leaders may be implemented, so that it's clear which marker the data component applies to. Toggle on "Automatic Leaders" if this feature is desired.

Map makers tend to include a slight text rotation on the data components to make the map more readable. 30 degrees is a popular point marker and data component rotation.

The following dialogue box displays the "Preferences" tab in the "Point Settings" dialogue box.

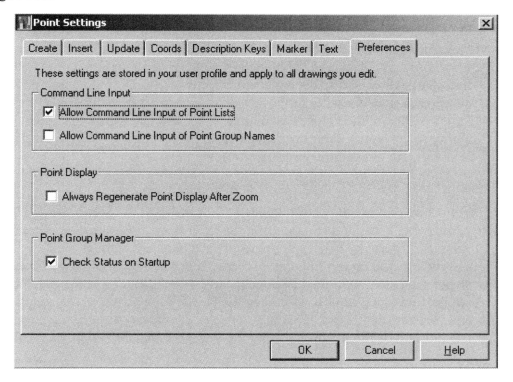

The "Command Line Input" area allows users to specify whether they want to type lists of points at the command line, or go directly into a dialogue box. Users which specify "Command Line Input" also have the ability to use dialogue boxes.

Points having their size dictated as a percentage of screen size can be automatically regenerated after changing the points zoom ratio in the drawing environment, but this can be memory intensive at times. Since the preferred method of setting up points is to implement a fixed height, this setting is not critical and is better left toggled off. A simple Regen (RE) will display your points when you want to see them after performing operations which will affect their appearance.

Now that we are all experts in setting up our project points, let's get down to business. You are working for a survey firm whose services have been retained to subdivide a residential parcel. You know that an analysis of the property will be required, so prior to surveying the site, you research the parcel and find in the deed that pipes were set which reflect the corners of the parcel. The developer has informed you that trees on the parcel being subdivided will be protected. As head of the survey department you instruct your crew to record field information which will allow you to locate every tree and iron pipe on the site.

In analyzing the field survey data upon your field crew's return, to your surprise, the field crew was able to locate all of the iron pipes on the site. Unfortunately, the field crew's data collector has malfunctioned and the only means of recording data was in a field book, so project points will need to be created in our database and drawing environments manually.

Exercise Instructions

- Logon to your workstation and begin a session of Land Desktop. If Land Desktop has been configured on your workstation to display the "Start Up" dialogue box, then cancel this feature so that the AutoCAD model space environment is displayed. If launching Land Desktop brings you directly into the AutoCAD model space environment, then Land Desktop has been configured to begin without the startup dialogue. If the AutoCAD Map "Task Pane" is shown, then close this dialogue box as well. Users that have the ability to customize their profiles can turn these features off permanently by following the procedure described in Assignment #1.

- Begin a new drawing by selecting "File→New". Choose to create a new project by picking the button labeled "Create Project". Select the "Prototype:" "Default (Feet)" from the drop down list. Give the project the "Name: " "Assign03-##" where the "##" reflects the number which was assigned to you on the first day of class. If the number assigned to you is a single digit (for example 4 as opposed to 14), then enter a zero preceding the single digit (such as 04). Write "Setting AEC Points" in the "Description:" and "Keywords:" areas. This will serve as a project summary which may be used to find or filter projects using the Project Manager.

- Select "OK" to bring you back to the "New Drawing" dialogue box and type the "Name:" "Assign03-##" where the "##" reflects the number which was assigned to you on the first day of class. Make sure that the "Project Name:" displays the correct project.

- Select the "Acad.dwt" template and then "OK" to generate a new drawing in the "DWG" folder of the current project. After selecting the "OK" button, you may be prompted to save changes to the previous drawing session. Since there were no objects in the previous drawing session worth saving, choose not to save changes.

- You will be prompted with the "Load Settings" dialogue box which displays a list of drawing setups to choose from. These setups are saved back to the path determined by the Network Administrator who installed the software. The default path is to the local machine and is displayed above. We will set up our parameters manually, so select "Next" to set up your project with the following parameters:

Units Area
> Linear Units = Feet
> Angle Units = Degrees
> Angle Display Style = Bearings
> Display Precision Linear = 2
> Display Precision Elevation = 2
> Display Precision Coordinate = 5
> Display Precision Angular = 4
> "Next"

Scale Area
> Horizontal = 100
> Vertical = 1
> Paper Size = 8 x 11(A)
> "Next"

Zone Area
> "Next"

Orientation Area
> "Next"

Text Style Area
> Leroy.stp
> L100

- Then select "Finish" and a screen will display providing the user with a summary of the settings chosen. Review the settings and select "OK". Although users have the ability to save settings for retrieval with future projects, assignments in this text require that you set up the parameters manually for practice in each project.

- You will be prompted with the "Create Point Database" dialogue box. Accept the default settings by choosing "OK".

- Create several new layers in the following manner:

Layer Name	Color	Linetype
C-ANNO-VPRT	White	Continuous
C-BNDY-LOTS	Magenta	Phantom
C-BNDY-PIPE	Red	Continuous
C-PNTS-DIVI	Yellow	Continuous
C-PNTS-SIDE	Cyan	Continuous
C-PNTS-TRAV	Red	Continuous
C-TREE-TRNK	Green	Continuous

- Select "Points→Point Settings" and Set up your points as displayed in the graphics at the beginning of this chapter.

- The first three points are traverse points. We will set the first two points manually. While on the layer C-PNTS-TRAV set points 1-3 using the command "Points→Create Points→Manual". You will have to enter (X,Y) coordinate pairs at the command line while executing this routine. Consider the relationship between (X,Y) and Northings & Eastings as you create these points.

Point	Northing	Easting	Elevation	Description
1	-34.8213	8.3773	128.00	BM
2	138.3838	108.3773	130.00	TRAV

- Set the third point, also being a traverse point using the command "Points→Create Points→Northing/Easting". You will be prompted for Northing and Easting coordinates separately. Notice the difference in syntax between this command and the previous command used to set points.

Point	Northing	Easting	Elevation	Description
3	69.9797	296.3158	132.00	TRAV

- Points 4-15 are side shots from our control traverse. While on the layer C-PNTS-SIDE, set points 4-15 using the command "Points→Create Points→Turned Angle". Implement the points option by typing "PO". If the software prompts you for a "Starting point", then Land Desktop is requesting that you use your "Node" object snap to identify points. If the software prompts you for a "Point number", then Land Desktop is requesting that you type in the point number as opposed to picking the point on the screen. The prompt can be changed back and forth while in the Land Desktop command using the toggle ".P". The occupied (Starting) point for shots 4-13 is 2. The backsight (Ending point) for points 4-13 is 1. The occupied point for shots 14-15 is 3. The backsight for points 14-15 is 2. Use the "Turned" option when setting these points.

Point	Angle Rt.	Distance	Description	Elevation
4	19.0121	108.46	TREE	128.50
5	125.4641	158.99	TREE	134.00
6	185.1842	146.41	TREE	133.50
7	224.1704	69.82	TREE	131.00
8	237.4337	181.84	TREE	133.50
9	315.0535	210.47	TREE	129.00
10	45.0000	240.42	IP	130.00
11	112.3333	184.09	IP	134.50
12	167.5835	272.67	IP	135.00
13	247.2627	184.09	IP	134.00
14	120.2339	247.47	IP	136.00
15	287.3008	206.50	IP	129.50

- We wish to begin subdividing our parcel. In doing so we would like to break the segment between points 10 and 11 into three equal pieces. Change to the layer "C-PNTS-DIVI". Select "Points→Point Settings→Create" and toggle on the Automatic Descriptions. Type in the description "DIVIDE". Toggle off the elevations by selecting the "None" option, since they will not be necessary or correct. Select "OK" to exit the "Point Settings" dialogue box.

- Use the command "Points→Create Points→Divide Object" to generate 4 points spaced equally between points 10 and 11. In order to accomplish this, select "PO" to designate that you would like to divide the segment using points as the input. For "First point" pick point "10" using your object snap. If the software is prompting you for a point number, you may type ".P" to select the "First point" using your object snap. For "Second point", using your object snap choose point "11". When prompted for the "Last point" press the "Enter" key on your keyboard. When prompted for the number of segments, type "3". Specify an offset of "0". You have now set 4 points identifying three segments.

- Change to the layer "C-BNDY-LOTS", turn off the layer "C-PNTS-DIVI," and draw a polyline between nodes 10,11,12,14,13,15 and back to 10 by typing "PL" at the command prompt and using your object snap. Make sure that your boundary matches the boundary in the picture in the reference figure at the end of this assignment. Edit the polyline using the "Pedit" command and turn "Ltgen" "On". This will allow the segments in your boundary to appear uniform.

- Change to the layer "C-BNDY-PIPE" and use the "Survey" Palette in the "Utilities→Symbol Manager" to insert an "Iron pin" symbol in the six locations where your crew picked up pipes. Notice that you have the option to select a node or to toggle on and off the option of setting the block at a particular point number by typing ".P". I suggest that you try both options.

- Change to the layer "C-TREE-TRNK" and use the "Plants" Palette in the "Utilities →Symbol Manager" to insert the "Tree 9" symbol where your crew picked up trees.

- Pick the layout tab currently titled "Layout1" so that this layout becomes active. Right click on this tab and choose to rename the layout tab to "Plat" Now that you are in paperspace, change to the layer "C-ANNO-VPRT" and create a landscape viewport 11'×8.5' having a scale of 1in=100ft (100SC). Make sure that your "ltscale" and "psltscale" variables are set appropriately for paperspace plotting as discussed in previous assignments.

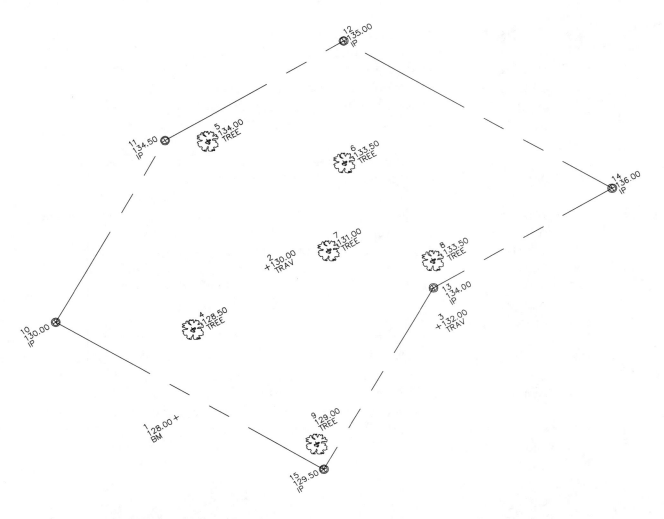

Reference Figure: Assignment 3

Assignment #4

Point Groups and the Land Desktop Project Manager

Recommended Assignments Prior to Working this Assignment:

Assignments 1-3

Required Assignments Prior to Working this Assignment:

Assignment 3

Goals and Objectives

The Land Desktop "Project Manager" is a fairly powerful utility which allows users to rename, create, copy, search, filter and delete projects. Users also have the opportunity to see which users have external database files open by selecting the "File Locks" option. A copy of the Land Desktop Project Manager is shown below for reference.

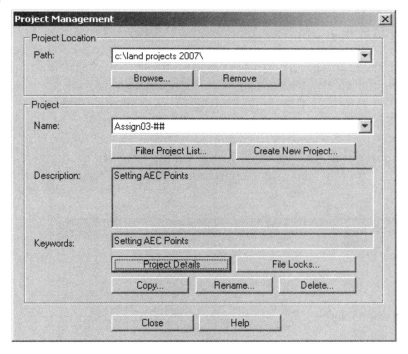

Land Desktop also provides a manager which allows users to organize their points in groups. Point grouping is a handy tool, because many of the Land Desktop commands allow users to build selection sets by choosing point groups. Users have the ability to export, erase, copy, rotate and translate points by merely informing Land Desktop which point group they want to perform the operation on. Don't let point grouping be a substitute for layer management. Good layer management is also important. In a well set up project, the two should complement each other.

Your design firm has been retained to prepare a plot plan for a residential subdivision which was built approximately a year ago. After conducting some research, you find that the original subdivision work was prepared by the surveying department in your design firm, but your client, the owner of the lot being developed is not the original subdivision developer. Therefore, you decide that you would like to start with your original project survey control, but you prefer to keep the project data files separate. In order to accomplish this, you decide to copy the project files using the Land Desktop "Project Manager" to generate a new project. After copying the project files, you rationalize that it would be wise to maintain the external database files, but to erase the copied AutoCAD drawings in the newly created project, so that other employees don't accidentally confuse the projects and access the wrong drawing file.

Exercise Instructions

- Logon to your workstation and begin a session of Land Desktop. If Land Desktop has been configured on your workstation to display the "Start Up" dialogue box, then cancel this feature so that the AutoCAD model space environment is displayed. If launching Land Desktop brings you directly into the AutoCAD model space environment, then Land Desktop has been configured to begin without the startup dialogue. If the AutoCAD Map "Task Pane" is shown, then close this dialogue box as well. Users that have the ability to customize their profiles can turn these features off permanently by following the procedure described in Assignment #1.

- Use the Land Desktop project manager (Projects→Project Manager) to copy "Assign03-##" to a new project named "Assign04-##" where the "##" reflects the number which was assigned to you on the first day of class. You can accomplish this by changing the project "Name:" to Assign03-## and choosing the "Copy" option.

- In the "Copy Project To" "Name:" location type "Assign04-##" where the "##" reflects the number which was assigned to you on the first day of class. If the number assigned to you is a single digit (for example 4 as opposed to 14), then enter a zero preceding the single digit (such as 04). In the "Description:" and "Keywords:" areas, enter the text "Point Groups". Choose "OK" and Land Desktop will ask you if you want to reassociate the copied drawing files so that the copied drawings reference the copied project. Choose "Yes" and then "Close."

- We now have a drawing file named "Assign03-##" in our "Assign04-##" project. It is important that we delete this file before proceeding so that there is no confusion as to which file is the correct "Assign03-##" drawing file. Use Windows Explorer or some other file manager to delete the "Assign03-##.dwg" drawing which resides in your "Assign04-##" project.

- Begin a new drawing by selecting "File→New" to bring up the "New Drawing" dialogue box. Type the "Name:" "Assign04-##" where the "##" reflects the number which was assigned to you on the first day of class.

- We are presently working with Project: "Assign04-##". Make sure that the "Project Name:" displays the correct project, select the "Acad.dwt" template and then "OK" to generate a new drawing in the "DWG" folder of the current project. After selecting the "OK" button, you may be prompted to save changes to the previous drawing session. Since there were no objects in the previous drawing session worth saving, choose not to save changes.

- You will be prompted with the "Load Settings" dialogue box which displays a list of drawing setups to choose from. These setups are saved back to the path determined by the Network Administrator who installed the software. The default path is to the local machine and is displayed above. We will set up our parameters manually, so select "Next" to set up your project with the following parameters:

    Units Area
        Linear Units = Feet
        Angle Units = Degrees
        Angle Display Style = Bearings
        Display Precision Linear = 2
        Display Precision Elevation = 2
        Display Precision Coordinate = 5
        Display Precision Angular = 4
        "Next"

Scale Area
  Horizontal = 100
  Vertical = 1
  Paper Size = 8 x 11(A)
  "Next"

Zone Area
  "Next"

Orientation Area
  "Next"

Text Style Area
  Leroy.stp
  L100

- Then select "Finish" and a screen will display providing the user with a summary of the settings chosen. Review the settings and select "OK." Although users have the ability to save settings for retrieval with future projects, assignments in this text require that you set up the parameters manually for practice in each project.

- You will not be prompted with the "Create Point Database" dialogue box because point information was copied from the "Assign03-##" project.

- Create new layers in the following manner:

Layer Name	Color	Linetype
C-ANNO-VPRT	231	Continuous
C-PNTS-SIDE-PIPE	Red	Continuous
C-PNTS-SIDE-TREE	Green	Continuous

- Set your point settings as you did at the beginning of Assignment 3. You will not be able to set the current point number to "1," because this point is all ready occupied. Instead, your point number should be set to "20."

- We will next set up point groups for the various point descriptions. Select the menu "Points→Point Management→Point Group Manager". The "Point Group Manager" will display as shown in the graphic below.

Point Groups

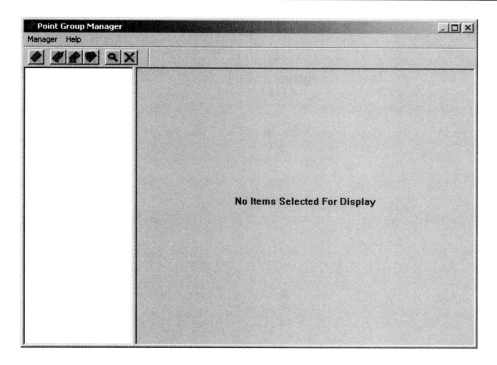

- The default Land Desktop 2007 installation creates pre-defined point groups for new projects if the "_CO-ENG(imperial) " prototype is used. The pre-defined point groups are intended to work with pre-defined "Description Keys" which will be discussed in later assignments. The default Land Desktop installation in previous releases used the "Default (Feet)" prototype and did not create these point groups.

- If point groups are present in the left pane, as shown in the graphic above, then delete them. Point groups may be deleted one at a time by selecting each of the point groups and picking the delete icon which appears as a red "X" in the toolbar just above the point group names.

- Select the green button in the upper left hand corner. This is the toolbar for creating a new point group. Give the point group the name "Tree" and select the tab named "Include." Toggle on the check box adjacent to the words "With Raw Description Matching" and type in the description "TREE" to the right. Keep in mind that by default point queries are case sensitive. "TREE" and "Tree" are not the same. This option can be turned off by unselecting the check box at the top of the dialogue box. Select "OK" to build your point group.

- Autodesk has implemented many menus in the Land Desktop software which may be activated with the right mouse button. While back in the Point Group Manger, right click on the newly created point group "TREE." If you needed to edit the point group, you could select the "Properties" option. Briefly observe the options in the right click menu and then choose the "Properties" option.

- Exit the "Point Group Properties" dialogue box, pick your point group and select the menu "Manager→Print to file" option to save the point group to the "Survey" directory your corresponding project. Name the file TREE.txt. This text file may be opened in any word processing editor for printing.

- Create point groups to include points with the following descriptions:

    BM
    DIVIDE
    IP
    TRAV
    TREE

- Print each of the point groups to a file in the survey directory of the corresponding project the same way that you printed the "Tree" group to a file.

- While on the "C-PNTS-SIDE-PIPE" layer, in the "Model" environment, use the command "Points→Insert Points to Drawing→Group" to insert the IPs to the current drawing.

- While on the "C-PNTS-SIDE-TREE" layer use the command "Points→Insert Points to drawing→Group" to insert the TREEs to the current drawing.

- Pick the layout tab currently titled "Layout1" so that this layout becomes active. Right click on the this tab and choose to rename the layout tab to "Plat." Now that you are in paperspace, change to the layer "C-ANNO-VPRT" and create a landscape viewport 11'×8.5' at a scale of 100. Make sure that your "ltscale" and "psltscale" variables are set appropriately for paperspace plotting.

- Save and exit this drawing. You have successfully completed this assignment.

Assignment #5

Importing and Exporting AEC Points

Recommended Assignments Prior to Working this Assignment:

Assignments 1-4

Required Assignments Prior to Working this Assignment:

None

Goals and Objectives

One of the most common ways for land surveyors to locate field information is through the implementation of a device known as a total station. A total station has the ability to store data which may contain information regarding the horizontal location, elevation and/or brief description of the data in question. Land Desktop gives users the ability to extract this data and import the data into a CAD environment.

There are a few things that Land Desktop users should be familiar with before importing data into the CAD environment. There are a variety of formats available for importing and exporting AEC Point data. Users also have the ability to create their own formats using the format manager.

There are a number of import options that a user should set up prior to importing data to the CAD environment. Import options are located in the Land Desktop menu "Points→Import/Export Points→Import Options".

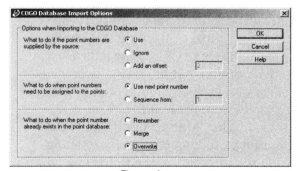
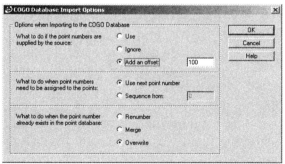

Case A   Case B

Although there are a variety of import options, the preceding dialogue boxes indicate two of the more desirable import scenarios which a user might choose prior to importing points from a text file. There is a table which contains the various scenarios and a brief description of the import options at the end of this chapter. In order to explain the differences between the various options in the "Import Options" dialogue box, consider the following example.

You currently have a project that consists of points 2-5. Your field crew has recently given you a text file that contains points 1-50 and the points in the text file do not match your project points.

The "**Use**" option suggests that a user would prefer to use the point numbers in the text file as opposed to assigning them a new point number.

The "**Ignore**" option suggests that a user would prefer to NOT use the point numbers in the text file and allow Land Desktop to renumber the text file point according to the other import settings.

If points are to be imported from a text file, and the corresponding point number is all ready occupied in the project, then it might be a good idea to import the points with an additive factor. This is what the "**Add an Offset**" accomplishes. An added offset of 100 would bring points 1-50 in as 101-150. If, for any reason, a user needed to go back to the text file, this would make it easier to track points between the text file and project points than some of the other import options.

The "**Renumber**" option will automatically assign new point numbers to the text file points if the corresponding point is all ready occupied in the project database. This option is a bit dangerous, because users who are not careful could conceivably import their text file and have it renumbered unsequentially to various unoccupied point ranges in the project database. If a user were to import points 1-50 from a text file using the "Renumber" option into a database that all ready occupied points 2-5, then the following could occur depending on some of the other import settings.

Text File	Project Database
Point 1 ------------------------->	Point 1
Points 2-5 ------------------------->	Points 6-9
Points 6-50 ------------------------->	Points 10-54

The "**Overwrite**" option will overwrite the points in the database with the points in the text file. If this option is used, the user must be careful to not overwrite project points with text file points having different data values. Most of the time this option is used, the duplicate points are control points and have the same coordinate data anyway.

The "**Merge**" option works like the overwrite option, but unlike the overwrite option, Merge combines the AEC point data of the point all ready in the project with the AEC point data of the point being read into the project from the text file. Users upgrading from Softdesk 8 products should be aware. The "Merge" option has a new meaning in Land Desktop. In S8, the "Merge" option would not change the existing point information in the current database. As a quick example, see the following flow diagram:

Existing Project Point		Text File Point	Resulting Project Point from Merge
Number:	2	--------------> 2	------------------> 2
Northing:	15	--------------> 27	------------------> 27
Easting:	20	--------------> 33	------------------> 33
Elevation:	105	-->	105
Description:	IP	-->	IP

If you are importing points into a project and you also choose to bring the points into a point group, you may not get what you expect depending on the import options. For example, if the merge option is selected and the user chooses to skip duplicate point numbers, then the points in the project with corresponding point numbers will be included in the point group. As an example, if a user were to import points 1-50 from a text file using the "Merge" option into a project containing points 2-5, and the user chose to add the points to a point group, then the following would occur.

- Point 1 would be added to the project.

- The user would be prompted with a dialogue box requesting to replace the project point number 2 with the text file point number 2, or to skip the text file point number 2 and go on to the text file point number 3.

- If the user chose to skip the text file points then the project would consist of points 1-50. Points 1, 6-50 would contain data obtained from the text file and points 2-5 would contain data from the original project points 2-5.

- One might expect the point group to contain points 1, 6-50. However, the point group will contain points 1-50. If the points are erased from the project using the point group for choosing the points to erase, then more points will be erased from the project, than actually imported from the text file.

When the options "Ignore" or "Renumber" are used, one more condition needs to be specified. Users need to tell Land Desktop whether they want the software to use the "Current number" set in the point settings by specifying "**Use next point number**" or use the next specified sequential point number specified in the import options "**Sequence from**".

Four years have elapsed since your company subdivided a parcel for a residential subdivision off of Summerfield Avenue. A new developer is in the process of purchasing the subdivision and has retained your firm to prepare a topographic map meeting the requirements of ALTA/ACSM to obtain funding for the property. You sent your survey crew out in the field with a total station and it is your task to create the map while making use of the data which they have stored for you in the data collector. (The map being generated for this lesson does not actually reflect the requirements established for ALTA/ACSM surveys.)

Exercise Instructions

- Logon to your workstation and begin a session of Land Desktop. If Land Desktop has been configured on your workstation to display the "Start Up" dialogue box, then cancel this feature so that the AutoCAD model space environment is displayed. If launching Land Desktop brings you directly into the AutoCAD model space environment, then Land Desktop has been configured to begin without the startup dialogue. If the AutoCAD Map "Task Pane" is shown, then close this dialogue box as well. Users that have the

- ability to customize their profiles, can turn these features off permanently by following the procedure described in Assignment #1.

- Begin a new drawing by selecting "File→New". Choose to create a new project by picking the button labeled "Create Project". Select the "Prototype:" "Default (Feet)" from the drop down list. Give the project the "Name:" "Assign05-##" where the "##" reflects the number which was assigned to you on the first day of class. If the number assigned to you is a single digit (for example 4 as opposed to 14), then enter a zero preceding the single digit (such as 04). Write "Importing and Exporting AEC Points" in the "Description:" and "Keywords:" areas. This will serve as a project summary which may be used to find or filter projects using the Project Manager.

- Select "OK" to bring you back to the "New Drawing" dialogue box and type the "Name:" "Assign05-##" where the "##" reflects the number which was assigned to you on the first day of class. Make sure that the "Project Name:" displays the correct project.

- Select the "Acad.dwt" template and then "OK" to generate a new drawing in the "DWG" folder of the current project. After selecting the "OK" button, you may be prompted to save changes to the previous drawing session. Since there were no objects in the previous drawing session worth saving, choose not to save changes.

- You will be prompted with the "Load Settings" dialogue box which displays a list of drawing setups to choose from. These setups are saved back to the path determined by the Network Administrator who installed the software. The default path is to the local machine and is displayed in the dialogue box. We will set up our parameters manually, so select "Next" to set up your project with the following parameters:

    Units Area
        Linear Units = Feet
        Angle Units = Degrees
        Angle Display Style = Bearings
        Display Precision Linear = 2
        Display Precision Elevation = 2
        Display Precision Coordinate = 5
        Display Precision Angular = 4
        "Next"

    Scale Area
        Horizontal = 100
        Vertical = 1
        Paper Size = 8 x 11(A)
        "Next"

Zone Area
"Next"

Orientation Area
"Next"

Text Style Area
Leroy.stp
L100

- Then select "Finish" and a screen will display providing the user with a summary of the settings chosen. Review the settings and select "OK". Although users have the ability to save settings for retrieval with future projects, assignments in this text require that you set up the parameters manually for practice in each project.

- You will be prompted with the "Create Point Database" dialogue box. Accept the default settings by choosing "OK".

- Create several new layers in the following manner:

Layer Name	Color	Linetype
C-ANNO-VPRT	231	Continuous
C-BLDG-EXST	Cyan	Continuous
C-FENC-EXST	White	Hidden2
C-PNTS-BLDG	Red	Continuous
C-PNTS-BNDY	Red	Continuous
C-PNTS-BMRK	Green	Continuous
C-PNTS-DYLO	Red	Continuous
C-PNTS-EPAV	Red	Continuous
C-PNTS-FENC	Red	Continuous
C-PNTS-PIPE	Red	Continuous
C-PNTS-TREE	Red	Continuous
C-PROP-BNDY	Magenta	Phantom
C-ROAD-DYLO	Green	Continuous
C-ROAD-EPAV	Yellow	Continuous
C-TREE-EXST	Green	Continuous

- Go to "Points→Point Settings→Create Tab" and make sure that the check box for "Insert to Drawing as created" is **NOT** checked. We wish to only add points to the COGO point database at the present time. Set up the other point parameters as described in the beginning of assignment 3. Graphics of the dialogue boxes are shown below for your reference.

Introduction to Land Desktop 2007

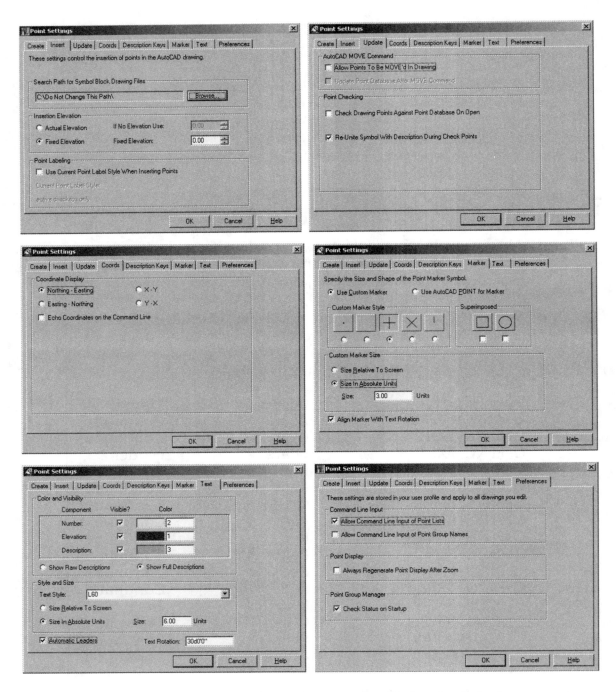

- Before we import our project points, we should also check the import options to ensure that the points come in the way that we would like them to. Select "Points→Import/Export Points→Import Options" and set up your options as shown in Case A earlier in this assignment.

- To import the points from a text file, run through the menu "Points→Import/Export Points→Import Points". Make sure that the format is "PNEZD (comma delimited)" for Point, Northing, Easting, Elevation, and Description. Select the manila folder to browse to the file which corresponds to this lesson. (Ask your instructor where lesson files are being stored on your school computer. Lesson files may be down loaded from www.schroff1.com.) Select the check box adjacent to "Add Points to Point Group" and pick the green button just below and to the right of the check box. This will give you the opportunity to add a point group. Type in "Assign05-Data" and choose "OK". Make sure that the point group that you just created is current and select "OK" again. The point group is current if it shows up in the white display just under the check box for adding the points to a point group. The import options will display again to give you the opportunity to review them prior to your import. Select "OK" to import your field points into the project point database.

- Before we insert points into the drawing environment from the COGO points database, make sure that you are in the "Model" environment. While on the layer "C-PNTS-BLDG", Select "Points→Insert Points to Drawing→Dialogue" to bring in the "Bldg" points. When the "Points to insert" dialogue box displays, click on the "Include" tab and toggle on the check box "With Raw Description Matching". Enter the description "bldg" just to the right. Keep in mind that by default the descriptions are case sensitive. This option can be toggled off above. Pick the button "Create Group" above to add your building points to a point group. Give the point group the name "BLDG" and select "OK". The group has been created. Select "OK" again to import your points.

- Now that your building points are in the drawing, while on the layer "C-BLDG-EXST", connect the points using lines as shown in the figure at the end of this assignment. After your linework is created turn on all layers except the building point layer. This will make the drawing easier to read as you bring in your next group of points.

- Change to the layer "C-PNTS-EPAV" and insert points to the drawing using the description "ep*". In this case the asterisk will allow a user to query all points which have descriptions that begin with the letters "ep". Without the asterisk, the query will only include points that have descriptions "ep". Don't forget to add these points to a point group. Change to the layer "C-ROAD-EPAV" and connect the linework using only lines and arcs. The following nomenclature is provided for your reference:

ep	Edge of Pavement
bc	Begin Curve
poc	Point on Curve
rc	Reverse Curve
cc	Compound Curve
ec	End Curve
bldg	Building
dylo	Double Yellow
f	fence

- Change to the layer "C-PNTS-DYLO" and insert points to the drawing using the description "dylo*". Don't forget to add these points to a group as you insert them to the drawing. A set of field notes indicates that the shot is at the center of the actual striping and that the stripes are 1ft apart. Make sure that your striping is on the layer "C-ROAD-DYLO" and that the stripes have been drawn correctly.

- Change to the layer "C-PNTS-PIPE" and insert points to the drawing using the description "IP". Don't forget to add these points to a group as you insert them to the drawing. Change to the Layer "C-PROP-BNDY" and connect the subdivision boundary as shown in the drawing at the end of this assignment.

- Change to the layer "C-PNTS-BNDY" and insert points to the drawing using the description "BNDY". Don't forget to add these points to a group as you insert them to the drawing.

- Change to the layer "C-PNTS-FENC" and insert points to the drawing by description "f". Don't forget to add these points to a group as you insert them to the drawing. Change to the layer "C-FENC-EXST" and draw in your fences. We will show fences on our topographical map on all property lines except the subdivision boundary. The crew picked up two shots for each of the fences on the three northern lots. Assume that the fences along the remaining 6 lots are all parallel to each other and are perpendicular to the straight segment in Eastside Court. Although the crew did not pick up fence alignment shots for the West lots, you remember setting points when these lots were subdivided several years ago. These fences may be drawn using the points having the "BNDY" description.

- Change to the layer "C-PNTS-TREE" and insert points to the drawing using the description "TREE". Don't forget to add these points to a group as you insert them to the drawing. Set the "C-TREE-EXST" layer current and use the Land Desktop command "Utilities→Symbol Manager" to insert the block "Tree 9" into the drawing at each of your tree locations. This tree block can be found under:

    Symbol Set: "Cogo Symbols"
    Category: "Cogo"
    Palette: "Plants"

    If you desire to type in point numbers in lieu of picking points in the AutoCAD environment using object snaps, type ".P" to toggle on the option to use "Point Numbers". A second ".P" will change the data entry option back to AutoCAD pick points.

- Change to the layer "C-PNTS-BMRK and insert points to the drawing using the description "BM".

- You now wish to export your points to a file in the format (P,X,Y,Z,D) so that a landscape architect can read them into a spreadsheet. Since this particular format does not exist, we will have to create one.

- Select "Points→Import/Export Points→Format Manager→Add→User Point File→OK" to bring up a dialogue box where we may set up our format. Name your format "PXYZD (Comma Delimited)". Change the extension to ".txt". Toggle on the "Delimited By" check box and make the delimiter a comma ",". Select the first unused database field (Column furthest to the left) and change it by scrolling down the list and choosing "Number". Select "OK" to exit. Go to the second unused database field and modify it to read the equivalent direction for X. Set the precision to "5" and select "OK" to exit. Go to the third unused database field and modify it to read the equivalent direction for Y. Set the precision to "5" and select "OK" to exit. Complete this process so that you have specified the equivalent for PXYZD. Elevation precision should be set to 2 decimal places. Select "OK" to add your new format to the format manager. This new format will now be available as an export option. Close the format Manager.

- Select "Points→ Import/Export Points→Export Points" to export your points to a new text file. Export in the format that you just created to the survey directory of the corresponding project with the name "Export.txt". Now open up the file in a session of notepad to view the file.

- Pick the layout tab currently titled "Layout1" so that this layout becomes active. Right click on this tab and choose to rename the layout tab to "Plat". Now that you are in paperspace, change to the layer "C-ANNO-VPRT" and create a landscape viewport 11′×8.5′ at a scale of 100.

- Turn on all of your linework as shown in the following picture. Save and exit this drawing. You have successfully completed this assignment.

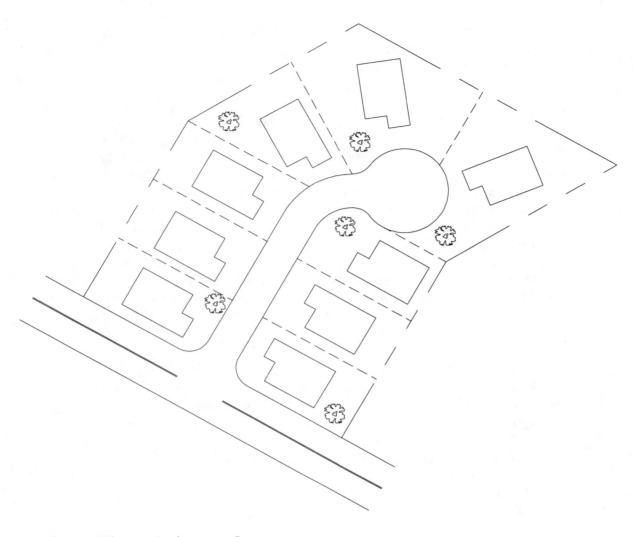

Reference Figure: Assignment 5

Importing and Exporting AEC Points

Matrix of options to select when importing points from a text file

Description	Pnt No. in Text File – Use	Pnt No. in Text File – Ignore	Pnt No. in Text File – Add Offset	IF Pnt No. Already in Project – Renumber	IF Pnt No. Already in Project – Merge	IF Pnt No. Already in Project – Overwrite	Use Next Pnt No.	Sequence From
Points are imported into project using text file point numbers and renumbered starting with the specified point in "Sequence From" if the point number in project is occupied.	X			X				X
Points are imported into project using the text file point numbers. Points in the project having common point numbers as those in the text file are automatically overwritten. Attribute data is merged.	X				X		N/A	N/A
Points are imported into project using the text file point numbers. Points in the project having common point numbers as those in the text file are automatically overwritten. Attribute Data is replaced.	X					X	N/A	N/A
Points are imported and renumbered starting with the current point number. If successive points in the project are occupied, then the point numbers are adjusted automatically to fill the voids in the project.		X		X			X	
Points are imported and renumbered starting with the "Sequential From" point number. If successive points in the project are occupied, then the point numbers are adjusted automatically to fill the voids in the project.		X		X				X
Points are imported into project using the current point number. Points in the project having common point numbers as those in the text file are automatically overwritten. Attribute data is merged.		X			X		X	
Points are imported into project using the "Sequential From" point number. Points in the project having common point numbers as those in the text file are automatically overwritten.		X			X			X

Matrix of options to select when importing points from a text file

	Pnt No. in Text File			IF Pnt No. All Ready in Project			Use Next Pnt No.	Sequence From
	Use	Ignore	Add Offset	Renumber	Merge	Overwrite		
Attribute data is merged.								
Points are imported and renumbered starting with the current point number. Points in the project having common point numbers as those in the text file are automatically overwritten. Attribute data is replaced.		X				X	X	
Points are imported and renumbered starting with the "Sequence From" point number. Points in the project having common point numbers as those in the text file are automatically overwritten. Attribute data is replaced.		X				X	X	X
Points are given a fixed offset. If successive renumbered points in the project are occupied, then the conflicting point numbers are adjusted sequentially starting with the current point number.			X	X			X	
Points are given a fixed offset. If successive renumbered points in the project are occupied, then the conflicting point numbers are adjusted sequentially starting with the "Sequential From" point number.			X	X			X	X
Points are given a fixed offset. Points in the project having common point numbers as those in the text file are automatically overwritten. Attribute data is merged.			X		X		N/A	N/A
Points are given a fixed offset. Points in the project having common point numbers as those renumbered in the text file are automatically overwritten. Attribute data is replaced.			X			X	N/A	N/A

Warning!! Point numbers are still added to point groups when using the merge option and a point is skipped.

Assignment #6

Rotation, Translation and Datum Adjustment

Recommended Assignments Prior to Working this Assignment:

Assignments 1-5

Required Assignments Prior to Working this Assignment:

Assignment 5

Goals and Objectives

Surveyors don't necessarily need to establish a **Datum** (Map reference elevation) and/or **Basis of Bearings** (Established bearing from which all reported map bearings are related) prior to mapping terrain. Field data can be adjusted after terrain is mapped. In addition, there are instances where a designer is looking for topographic features, and while these site features need to be located relative to each other, their orientation may not need to be tied down to the remainder of the world. In such cases, drawings do not need to be rotated to an established bearing or shifted to a known datum.

When Surveyors implement an assumed datum and/or basis of bearings, there are a couple of things to keep in mind. Surveyors should avoid implementing values for the assumed datum and/or basis of bearing that are reasonably close to the actual datum and/or basis of bearing. If a value is implemented that is obviously not correct, then the risk of someone accidentally using the value as a true datum or basis of bearing is reduced.

As an example, if a Surveyor is creating a topographic map of an area, for which the true elevation is in the neighborhood of 115ft, and it is the Surveyor's intent not to relate the survey to an established benchmark, then it would be wise for the Surveyor to assume a datum elevation drastically different than 115ft. If the Surveyor uses an assumed elevation of 100ft, then a map interpreter might make the mistake of believing that the topographic elevations are true elevations rather than assumed elevations. However, if the Surveyor uses an assumed elevation of 600ft, and there is not very much fall across the site, then the map interpreter is less likely to make the mistake of believing that the topographic elevations are true elevations.

You have just discovered that one of your junior technicians made a mistake when creating the control for a subdivision construction survey that you are working with. The City Datum that he looked up for you doesn't correlate with the map and is off by 1.25ft and you have discovered that the basis of bearings does not correspond with the map that you provided him. You also notice that some of the coordinates in the project point database have negative values. You recognize that it is a more difficult task to work with negative coordinates in the data collector and you anticipate sending the field crew back to the site in a couple of weeks for construction staking, so you would like to change this as well.

In the above scenario, you are presented with three tasks that you would like to accomplish. You want to alter the datum of the survey, rotate the survey to the correct basis of bearings and move the coordinates out into space to alleviate the possibility of negative coordinate values. You recognize that there are several ways to address this issue, but you want to select the best option.

Land Desktop has routines in the menu "Projects→Drawing Setup→Orientation" which will allow users to specify relationships to use between the Land Desktop database coordinate system and the AutoCAD world coordinate system. However, users often decide not to modify these settings for the following reasons:

1. Modifying these settings will not alter the Northings and Eastings of the points that have all ready been created in the database and we need to modify the Northings and Eastings so that the coordinates will not have negative values.

2. Altering these settings will create a situation in which the AutoCAD and Land Desktop coordinate systems are no longer similar. In other words, the world coordinate pair (0,0) would no longer match the Land Desktop coordinate pair Northing=0, Easting=0. This would limit users to Land Desktop commands when working in their project. Commands such as "List" and "ID" would no longer work the way a user might anticipate because of the phase shift in coordinate systems.

3. North may no longer be up in your drawing. This might make your job or someone else's more difficult since many users are more comfortable working with drawings where North is straight up.

There are commands available in the "Points" menu for rotating and translating the coordinate systems. These commands are "Points→Edit Points→Translate Points" and "Points→Edit Points→Rotate Points". Users often choose not to use either of these commands, because the project must be set up in "Single user mode" to access them and they yield similar results to items 2 and 3 stated above.

Points can be moved and rotated using the "Points→Edit Points→Move" and "Points→Edit Points→Rotate" commands. These are more desirable options since they alter points in both the project point database and the AutoCAD world coordinate system. Unfortunately, these commands will not have any impact on the linework which you likely spent hours creating.

After weighing the advantages and disadvantages of each scenario, many users decide to conduct a simple AutoCAD move and rotate. Users then update the project points database using a routine called "Check Points". The datum adjustment is approached with the Land Desktop "Datum Adjustment" command.

Rotation, Translation and Datum Adjustment

Exercise Instructions

- Logon to your workstation and begin a session of Land Desktop. If Land Desktop has been configured on your workstation to display the "Start up" dialogue box, then cancel this feature so that the AutoCAD model space environment is displayed. If launching Land Desktop brings you directly into the AutoCAD model space environment, then Land Desktop has been configured to begin without the startup dialogue. If the AutoCAD Map "Task Pane" is shown, then close this dialogue box as well. Users that have the ability to customize their profiles can turn these features off permanently by following the procedure described in Assignment #1.

- Begin a new drawing by selecting "File→New". Choose to create a new project by picking the button labeled "Create Project". Select the "Prototype:" "Default (Feet)" from the drop down list. Give the project the "Name:" "Assign06-##" where the "##" reflects the number which was assigned to you on the first day of class. If the number assigned to you is a single digit (for example 4 as opposed to 14), then enter a zero preceding the single digit (such as 04). Write "Rotation, Translation & Datum Adjustment" in the "Description:" and "Keywords:" areas. This will serve as a project summary which may be used to find or filter projects using the Project Manager.

- Select "OK" to bring you back to the "New Drawing" dialogue box and type the "Name:" "Assign06-##" where the "##" reflects the number which was assigned to you on the first day of class. Make sure that the "Project Name:" displays the correct project.

- Select the "Acad.dwt" template and then "OK" to generate a new drawing in the "DWG" folder of the current project. After selecting the "OK" button, you may be prompted to save changes to the previous drawing session. Since there were no objects in the previous drawing session worth saving, choose not to save changes.

- You will be prompted with the "Load Settings" dialogue box which displays a list of drawing setups to choose from. These setups are saved back to the path determined by the Network Administrator who installed the software. The default path is to the local machine and is displayed in the dialogue box. We will set up our parameters manually, so select "Next" to set up your project with the following parameters:

Units Area
    Linear Units = Feet
    Angle Units = Degrees
    Angle Display Style = Bearings
    Display Precision Linear = 2
    Display Precision Elevation = 2
    Display Precision Coordinate = 5
    Display Precision Angular = 4
    "Next"

Scale Area
    Horizontal = 100
    Vertical = 1
    Paper Size = 8 x 11(A)
    "Next"

Zone Area
    "Next"

Orientation Area
    "Next"

Text Style Area
    Leroy.stp
    L100

- Then select "Finish" and a screen will display providing the user with a summary of the settings chosen. Review the settings and select "OK". Although users have the ability to save settings for retrieval with future projects, assignments in this text require that you set up the parameters manually for practice in each project.

- You will be prompted with the "Create Point Database" dialogue box. Accept the default settings by choosing "OK".

- Insert "Assign05-##" using "Insert→Block→Browse" command at the coordinate pair (0,0) with a scale of 1 and rotation angle of 0. [If the "Insert" menu is not available, then change "Workspaces" by selecting the Workspace "Land Desktop Complete" from the drop down list in the "Workspaces" toolbar. If the "Workspaces" toolbar is not displayed on the screen, then it can be invoked using the command "Projects→Workspaces". ("Workspaces" is a new feature that was introduced in the AutoCAD 2006 suite of products and has taken the place of the "Menu Palette Manager" used in previous versions of Land Desktop.)] Explode the drawing one time so that AutoCAD and Land Desktop are able to recognize drawing entities.

- At this time, the external COGO points database does not contain point information. The drawing, however, does contain AEC COGO point information. We would like to have Land Desktop update the external database to include point data from the current drawing. To accomplish this select "Points→Check Points→Modify Project" to write your AutoCAD point object information to the project point database. Choose the options as shown in the graphic below.

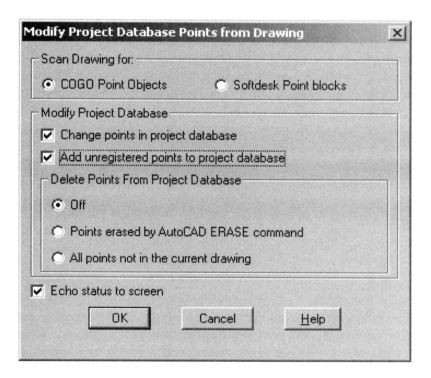

- We want to alter the descriptions for the trees to read "Oak". One way to alter the descriptions is through the use of the Land Desktop command "Edit Points". Select "Points→Edit Points→Edit Points" and select the radio button for "List All Points". Select the tab "Edit" and pick the column header "Raw Description" to sort the points by this field. Scroll down the list and change the descriptions of the "Tree" points to "OAK" by picking and editing them one at a time. (Descriptions can also be copied and pasted using the right click mouse option and the windows copy and paste macros Ctrl + C and Ctrl + V.) After changing the descriptions of all trees to read "OAK", select "OK" to exit the "Edit Points" dialogue box.

- We will next correct for the vertical error by adjusting the project datum upward 1.25ft. To accomplish this, select "Points→Edit Points→Datum", and type "1.25" when prompted for a change in elevation. Specify "All" of the project points when prompted for the range. (The project datum can be adjusted downward by entering a negative value for the adjustment.)

- We are going to use the AutoCAD "Move" command to translate the drawing entities out into space to eliminate the negative coordinate values. Make sure that all layers are turned on and thawed. Before we move these points, we want to make sure that the point settings are correct. Select "Points→Point Settings→Update" and set the parameters as identified below:

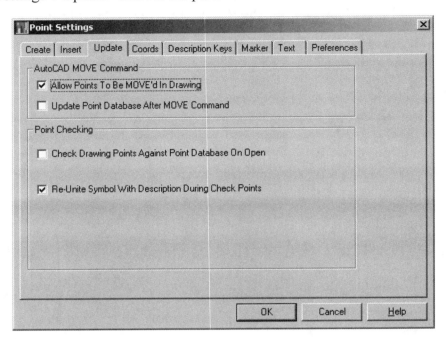

- Land Desktop may remember that the points set in the drawing environment from Assignment 5 were once configured to prevent a move. While this setting can be toggled back on in the current drawing to allow point moving using AutoCAD commands, simply toggling the setting on to permit a move may not allow points which were blocked into a drawing to be moved. If this occurs, simply toggle on the setting to allow moving, exit the dialogue box and perform an AutoCAD "Regeneration all" (by typing "REA" at the command line). Enter the dialogue box again to toggle the setting off. Exit the dialogue box and perform an AutoCAD "Regeneration all" (by typing "REA" at the command line). Enter the dialogue box on last time to toggle the setting back on again. This will cache the point settings and allow users to move points using AutoCAD commands.

- Select "Modify→Move" (or by typing "M" at the command line) and type "All" when prompted to select entities. Use your object snap to select the node for the point which represents the benchmark and move it to the absolute coordinate of (1000,1000).

- We now wish to rotate our drawing so that its bearings match the bearings of our recorded map. Using the AutoCAD "Rotate" command in the "Modify" menu (or by typing "Ro" at the command line), type "All" and elect to rotate all entities about the node for the point which represents the project benchmark. We find from our record map that the angle of the line made between pipes at points 15 and 10 is due West. Choose the "Reference" option in the AutoCAD rotate command, select the first pick point at node 15 and the second pick point at node 10. Our new angle is due west. The AutoCAD angle which corresponds to due West is 180 degrees. If desired, rather than keying in this value, users can use AutoCAD's "Ortho" feature.

- Now that we have moved and rotated our AutoCAD entities, our AutoCAD coordinates do not match our project point database. Our AutoCAD coordinates are the ones that we want to maintain. Check and see for yourself. ID the AutoCAD entity which corresponds to the benchmark and you will see that the benchmark, as expected is located at the coordinate pair (1000,1000). Select "Points→List Points" to see the Land Desktop database point location. You should see that the coordinates are different. Another way to compare the X/Y coordinates to Northing/Easting coordinates is to use the command "Inquiry→North/East". Typing ".P" at the command line will allow users to toggle back and forth between entering "Point Numbers" (Land Desktop Points) and using AutoCAD object snaps to pick points (Not necessarily Land Desktop points) in the drawing environment.

- We wish to update our Land Desktop coordinate database. In order to accomplish this select "Points→Check Points→Modify Project→OK". The options should be set as in the previous "Check Points".

- As a final task export your project points to a text file named "Export" in the survey directory for this project with the format (PNEZD) using a comma delimiter. Review the previous assignment if you can't remember how to accomplish this.

- Pick the layout tab currently titled "Layout1" so that this layout becomes active. Right click on this tab and choose to rename the layout tab to "Plat". Now that you are in paperspace, change to the layer "C-ANNO-VPRT" and create a landscape viewport 11'×8.5' at a scale of 100.

- Save and exit this drawing. You have successfully completed this assignment.

NOTES:

Assignment #7
Lines and Curves

Recommended Assignments Prior to Working this Assignment:

Assignments 1-6

Required Assignments Prior to Working this Assignment:

None

Goals and Objectives

A parcel map is one area with which technicians in the Civil Engineering and Land Surveying fields should be familiar. Although parcel maps have been used historically for various objectives, and although there are exceptions, they are presently used to subdivide land into four parcels or fewer. Information which may be found on a parcel map includes street right of way, boundary information, and/or easements. Final maps are similar to Parcel Maps because they are also used to subdivide land. Although there are exceptions, Final Maps are more commonly used in present day to subdivide parcels into five lots or more.

Land Desktop contains a variety of line and curve utilities which aide in the creation of such maps. In this assignment you will create linework for a final map while implementing a few of these utilities.

Exercise Instructions

- Logon to your workstation and begin a session of Land Desktop. If Land Desktop has been configured on your workstation to display the "Start Up" dialogue box, then cancel this feature so that the AutoCAD model space environment is displayed. If launching Land Desktop brings you directly into the AutoCAD model space environment, then Land Desktop has been configured to begin without the startup dialogue. If the AutoCAD Map "Task Pane" is shown, then close this dialogue box as well. Users that have the ability to customize their profiles can turn these features off permanently by following the procedure described in Assignment #1.

- Begin a new drawing by selecting "File→New". Choose to create a new project by picking the button labeled "Create Project". Select the "Prototype:" "Default (Feet)" from the drop down list. Give the project the "Name:" "Assign07-##" where the "##" reflects the number which was assigned to you on the first day of class. If the number assigned to you is a single digit (for example 4 as opposed to 14), then enter a zero preceding the single digit (such as 04). Write "Lines and Curves" in the "Description:" and

"Keywords:" areas. This will serve as a project summary which may be used to find or filter projects using the Project Manager.

- Select "OK" to bring you back to the "New Drawing" dialogue box and type the "Name:" "Assign07-##" where the "##" reflects the number which was assigned to you on the first day of class. Make sure that the "Project Name:" displays the correct project.

- Select the "Acad.dwt" template and then "OK" to generate a new drawing in the "DWG" folder of the current project. After selecting the "OK" button, you may be prompted to save changes to the previous drawing session. Since there were no objects in the previous drawing session worth saving, choose not to save changes.

- You will be prompted with the "Load Settings" dialogue box which displays a list of drawing setups to choose from. These setups are saved back to the path determined by the Network Administrator who installed the software. The default path is to the local machine and is displayed in the dialogue box. We will set up our parameters manually, so select "Next" to set up your project with the following parameters:

  <u>Units Area</u>
  Linear Units = Feet
  Angle Units = Degrees
  Angle Display Style = Bearings
  Display Precision Linear = 2
  Display Precision Elevation = 2
  Display Precision Coordinate = 5
  Display Precision Angular = 4
  "Next"

  <u>Scale Area</u>
  Horizontal = 60
  Vertical = 1
  Paper Size = 8 x 11(A)
  "Next"

  <u>Zone Area</u>
  "Next"

  <u>Orientation Area</u>
  "Next"

  <u>Text Style Area</u>
  Leroy.stp
  L80

- Then select "Finish" and a screen will display providing the user with a summary of the settings chosen. Review the settings and select "OK". Although users have the ability to save settings for retrieval with future projects, assignments in this text require that you set up the parameters manually for practice in each project.

- You will be prompted with the "Create Point Database" dialogue box. Accept the default settings by choosing "OK".

- Create several new layers with the following parameters:

Layer	Color	Linetype
C-ANNO-VPRT	White	Continuous
C-PROP-BSLN	Green	Dashed2
C-PROP-LINE	Magenta	Phantom
C-PNTS-TOPO	Red	Continuous

- Set your point settings (Refer to Assignment 3 if you can't remember how this was accomplished) so that your point text comes in with the L80 text style. Elect to turn off the "Elevation" and "Description" data components. Using the "+" marker, your marker height should be set at ½ the model space text height. The corresponding marker height for a text height of 4.8 units is 2.4. Set a text rotation angle of 30 degrees.

- While on the layer "C-PNTS-TOPO", select "Points→Import/Export Points→Import Points" using the "PNEZD" comma delimited format to read in points from the file which corresponds to this lesson. Lesson files may be downloaded from www.schroff1.com. As you read these points in, add them to a new point group named "All_Points". These points represent the starting vertices of the boundary that you are about to create. Zoom Extents to view your points.

- Change to the layer "C-PROP-LINE" and use the Land Desktop command "Lines/Curves→By Point # Range" to connect spot shots 1-7. This may be accomplished by typing the range "1-7" at the command line and then "Enter". "Enter" again to toggle out of the command.

- Once these lines are drawn, you can use the "Lines/Curves→By Direction" command to work your way around the boundary entering line information as shown in the figure at the end of this chapter.

  Upon entering the command, you will be prompted for a "Starting Point". Recognize that "Starting Point" and "Point Number" are not the same thing. If Land Desktop is requesting a "Point Number", then typing ".P" will toggle the prompt back to "Starting Point". Use your object snap to select the point which corresponds to point number 7. Land Desktop will next prompt for a quadrant. Quadrants are numbered 1-4 and have the following correlation:

```
4-NW | 1-NE
-----|-----
3-SW | 2-SE
```

Enter the quadrant number "3" which corresponds to the southwest direction. Land Desktop will next ask for the bearing. The bearing format is "Degrees.MinutesSeconds". Enter the angle "21.3447". Land Desktop will then prompt for the distance. Enter the length of the line as "113.31". Continue creating lines using the same procedure as is outlined above.

- When you get to the curves, you will need to use the commands "Lines/Curves→From End of Object" and "Lines/Curves→Reverse or Compound" in conjunction with the "Length" to create these curves.

   Upon entering the "Lines/Curves→From End of Object" command, you will be asked to select the starting entity. Use your "Nearest" object snap to pick the line at the end closest to where you want the curve to commence. When prompted, type "R" for Radius and enter a radius value. Keep in mind that the value of the radius dictates the direction that the curve breaks. A negative radius value will force the curve to break to the left while a positive radius value forces the curve to break to the right. As an example, if you want the curve to break to the left with a 25ft radius, enter a value of "-25". If you want the curve to break to the right, enter a value of "25". When prompted, elect to enter the curve length by typing "L", pressing "Enter" and then entering the appropriate length.

   Upon entering the "Lines/Curves→Reverse or Compound" command, when prompted, select the starting curve nearest the end you want to create the reverse. Type "R" to request the reverse option. Enter the radius as a positive value. As an example enter "42" if a 42ft radius is desired. Type "L" for length and enter the appropriate length in feet when prompted. Another "Enter" will toggle you out of the command.

- The building setback lines are merely offsets from the boundary lines. They should be put on the "C-PROP-BSLN" layer. Building setback lines for this assignment are 20ft from the exterior boundary and 10ft from interior boundary lines.

- All of the linework on the assignment 7 figure located at the end of this chapter should be included in your drawing. The North arrow, text, dimensions and line labels have been included for your reference and are not required with this assignment. Do **NOT** label the lines in this assignment, because this assignment will be used later to create a line/curve tables and if you label the lines now, creating a table might be a more difficult task in future lessons.

- Pick the layout tab currently titled "Layout1" so that this layout becomes active. Right click on this tab and choose to rename the layout tab to "Plat". Now that you are in paperspace, change to the layer "C-ANNO-VPRT" and create a landscape viewport 11′×8.5′ at a 60 scale. Your drawing will not entirely fit in this viewport at this scale. The orientation must be altered. We will address this in a future project by implementing a "Dview Twist". In the mean time, center the drawing as best you can.

- Save and exit this drawing. You have successfully completed this assignment.

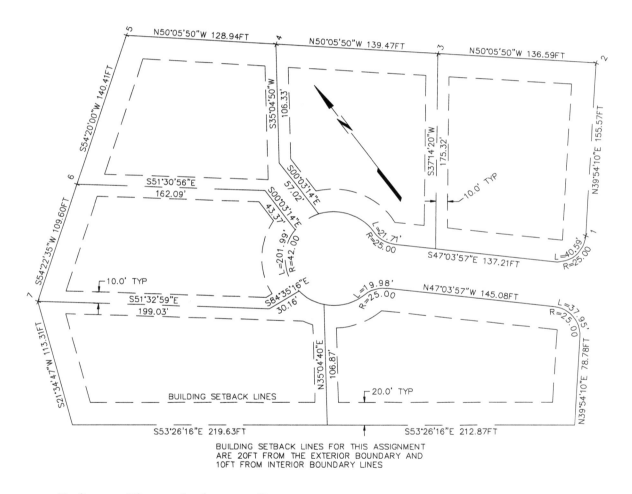

Reference Figure: Assignment 7

NOTES:

Assignment #8
Line Labeling

Recommended Assignments Prior to Working this Assignment:

Assignments 1-7

Required Assignments Prior to Working this Assignment:

Assignment 7

Goals and Objectives

      Line and curve labeling is one of the easiest ways to convey object geometry on a hard copy plot. Many Land Desktop users prefer Line/curve labeling to the traditional implementation of dimension strings because it's quick, easy to master and it gives users the ability to include a variety of information. One of the things that makes line/curve labeling so valuable is its ability to extract AutoCAD object information (such as Northings, Eastings, Direction, Radius and Length, etc.) which users typically include on the hard copy plot.

      Land Desktop contains a variety of line and curve label styles. The software is also customizable so that users have the ability to create their own line and curve label styles. Formulas may be implemented by creative users who want to set up label styles for labeling slopes in vertical drawing views.

      It's important to understand that label styles are created and set current in two different menus. This allows users to select available label styles rapidly without being inconvenienced by sieving through all of the style parameters every time users want to annotate lines and curves. Label styles are created using the command "Labels→Edit Label Styles". The graphic below shows the "Edit Label Styles" dialogue box. Several examples are also shown in the pages following.

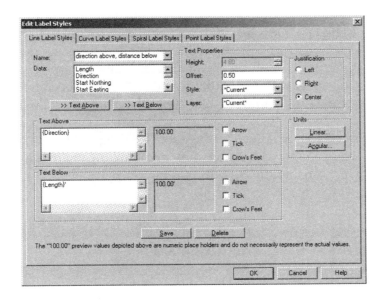

Label styles work similar to text styles in that users assign a name to a group of parameters. Although names can be called virtually anything, it's always best to generate names which will mean something to the next person that sits down at your workstation. As an example, the name "Direction Above, Distance Below" provides information that describes what the label style is going to accomplish more so than the name "Lisa". Because users are merely assigning a name to a group of parameters, it's important to remember that the label style may not generate the annotations which it describes if the label style hasn't been set up appropriately.

Annotations may be implemented above and/or below objects. If your objective is to generate a label style which merely annotates the length on top of linework, simply pick the "Data:" type which corresponds to the annotation you desire, in this case "Length" and select the "Text Above" button. The formula now appears in the "Text Above" area of the "Edit Label Styles" dialogue box.

Users have the ability to implement Arrows, Ticks and Crow's feet with their label styles. Below is an example of a label style having the Direction above and the Distance below. The label style also includes a directional "Arrow" and "Tick" marks.

$$N90°00'00"E$$
$$190.56'$$

Below is an example of a label style having the Direction and Distance above. This label style includes "Crow's Feet".

$$N\ 90°00'00"\ E\quad 190.56'$$

Users can assign text styles and the layer on which the annotations are placed, or allow the label style to make use of the "Current" AutoCAD "Text Style" and "Layer". The "Offset" value is not an absolute distance in feet. Instead, users are to specify a percentage of the text height. Users also have the ability to specify text justification and label precision by choosing the appropriate radio button and selecting the buttons specified in the "Units" area of the "Edit Label Styles" dialogue box.

The "Save" button would better be described as "Apply" because users are not required to pick this button unless they wish to apply the changes and remain in the dialogue box.

Label styles are saved by default to the path "C:\Program Files\Land Desktop 2007\Data\labels\". The path may be changed by invoking the "Settings" dialogue box "Labels→Settings" and choosing the Browse option. The "Labels Settings" dialogue box is shown on the following page for reference.

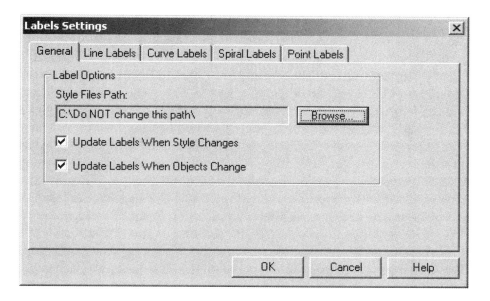

Labels can be applied Dynamically or Statically. Static labels are merely multiline text, and as such, they are not updatable. Check boxes are included on the "General" tab of the "Label Settings" dialogue box to allow Dynamic labels to update automatically every time users either:

– make adjustments to the label style in the "Edit Label Styles" dialogue box, or

– change an object's orientation and/or length in the drawing environment.

The following dialogue box shows the "Line Labels" tab of the "Label Settings" dialogue box.

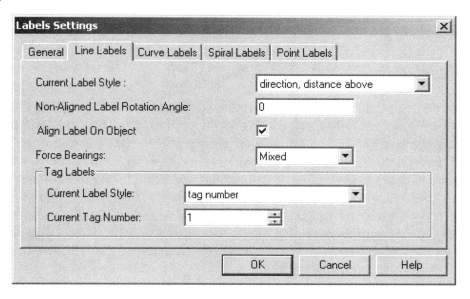

This is the dialogue box and tab that users would invoke when setting the current label style. Most users elect to align labels on the linework. Most drawings are

orientated so that the direction North is pointing upward on the plotted product. However, there is the occasional project where it is desirable to have North point downward. In these instances, Land Desktop gives users the ability to toggle off the "Align Label on Object" option and specify a "Non-Aligned Label Rotation Angle", so that the label text is no longer upside down.

The line made up having the bearing N30W has the same orientation as a line having the bearing S30E. Users can implement "Forced" bearings which will instruct the software to always label lines with reference to a fixed cardinal direction (North/South). Users can also implement "Mixed" bearings which will allow the cardinal direction (North/South) to be determined by the direction the line is traveling. The direction which lines travel is determined by the starting and ending points for the line. A line's direction can be changed by selecting the line, right clicking and choosing "Flip Direction".

The dialogue bar shown below may be invoked using the "Labels→Show Dialogue Bar" menu and is handy when you have a lot of labeling using multiple label styles. The drop down list gives users quick access the styles previously created using the "Edit Label Styles" dialogue box. The 4 buttons on the left give users quick access to Tag Styles, the "Edit Label Styles", the "Label Settings" and help dialogue boxes.

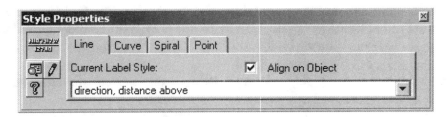

In this assignment, users will set up new line/curve label styles and label the final map excerpt from assignment 7. Although some of the label styles may have all ready been set up on the workstation at which you are sitting, you will need to view the label style to ensure that the parameters have been set up appropriately.

Exercise Instructions

- Logon to your workstation and begin a session of Land Desktop. If Land Desktop has been configured on your workstation to display the "Start Up" dialogue box, then cancel this feature so that the AutoCAD model space environment is displayed. If launching Land Desktop brings you directly into the AutoCAD model space environment, then Land Desktop has been configured to begin without the startup dialogue. If the AutoCAD Map "Task Pane" is shown, then close this dialogue box as well. Users that have the ability to customize their profiles can turn these features off permanently by following the procedure described in Assignment #1.

- Begin a new drawing by selecting "File→New". Choose to create a new project by picking the button labeled "Create Project". Select the "Prototype:" "Default (Feet)" from the drop down list. Give the project the "Name:" "Assign08-##" where the "##" reflects the number which was assigned to you on the first day of class. If the number assigned to you is a single digit (for example 4 as opposed to 14), then enter a zero preceding the single digit (such as 04). Write "Line Labeling" in the "Description:" and "Keywords:" areas. This will serve as a project summary which may be used to find or filter projects using the Project Manager.

- Select "OK" to bring you back to the "New Drawing" dialogue box and type the "Name:" "Assign08-##" where the "##" reflects the number which was assigned to you on the first day of class. Make sure that the "Project Name:" displays the correct project.

- Select the "Acad.dwt" template and then "OK" to generate a new drawing in the "DWG" folder of the current project. After selecting the "OK" button, you may be prompted to save changes to the previous drawing session. Since there were no objects in the previous drawing session worth saving, choose not to save changes.

- You will be prompted with the "Load Settings" dialogue box which displays a list of drawing setups to choose from. These setups are saved back to the path determined by the Network Administrator who installed the software. The default path is to the local machine and is displayed in the dialogue box. We will set up our parameters manually, so select "Next" to set up your project with the following parameters:

    Units Area
        Linear Units = Feet
        Angle Units = Degrees
        Angle Display Style = Bearings
        Display Precision Linear = 2
        Display Precision Elevation = 2
        Display Precision Coordinate = 5
        Display Precision Angular = 4
        "Next"

    Scale Area
        Horizontal = 60
        Vertical = 1
        Paper Size = 8 x 11(A)
        "Next"

    Zone Area
        "Next"

Orientation Area
"Next"

Text Style Area
Leroy.stp
L80

- Then select "Finish" and a screen will display providing the user with a summary of the settings chosen. Review the settings and select "OK". Although users have the ability to save settings for retrieval with future projects, assignments in this text require that you set up the parameters manually for practice in each project.

- You will be prompted with the "Create Point Database" dialogue box. Accept the default settings by choosing "OK".

- Create several new layers with the following parameters:

Layer	Color	Linetype
C-ANNO-DIMS	Cyan	Continuous
C-ANNO-VPRT	White	Continuous
C-BNDY-LABL	White	Continuous

- Insert "Assign07-##" using "Insert→Block→Browse" command at the coordinate pair (0,0) with a scale of 1 and rotation angle of 0. [If the "Insert" menu is not available, then change "Workspaces" by selecting the Workspace "Land Desktop Complete" from the drop down list in the "Workspaces" toolbar. If the "Workspaces" toolbar is not displayed on the screen, then it can be invoked using the command "Projects→Workspaces". ("Workspaces" is a new feature that was introduced in the AutoCAD 2006 suite of products and has taken the place of the "Menu Palette Manager" used in previous versions of Land Desktop.)] Zoom Extents and explode the drawing one time so that AutoCAD and Land Desktop are able to recognize the individual drawing objects.

- Select "Labels→Edit Label Styles" and proceed to set up three new label styles with the following names:

   1. Direction Above, Distance Below
   2. Direction, Distance, Bndy Above
   3. Direction, Distance, Bndy Below

- The following is an example of how the label style should be set up for line label style number 1:

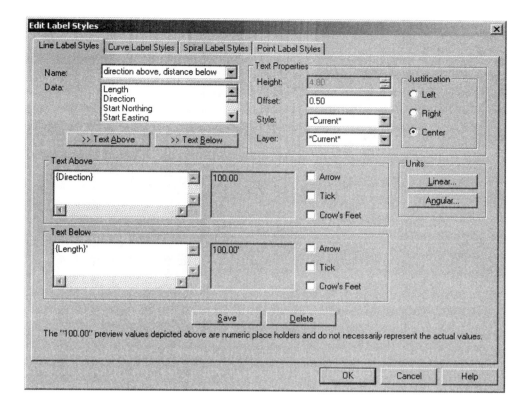

- The following is an example of how the label style should be set up for line label style number 2:

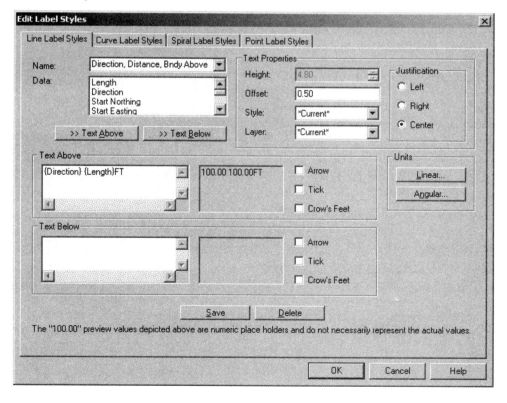

- Select "Labels→Settings" and set your label style under the "Line Labels" tab to label style 2 described earlier in this assignment.

- Set the text style in the AutoCAD "Text Style" dialogue box to an L80 set up for a 60 scale.

- Set the "C-BNDY-LABL" layer current and label the boundary line made up between points 2 and 3 in the same fashion that the boundary is labeled for this assignment. This can be accomplished by using the command "Labels→Add Dynamic Labels" and selecting the line which you want the label to appear. Keep in mind that the software perceives North as being above the line and South as being below the line. Using the label style "Direction, Distance, Bndy Above" will place the label on the North side of the line, whether you perceive the label to be above the line or not.

  The linework can also be labeled by selecting the line, right clicking and choosing the option "Add Dynamic Label".

  Experienced users often advise that they don't recommend altering the properties of individual objects, because doing so makes it difficult to control them at the parent level. The linetype is generally changed by layer so that all objects drawn on the subject layer can take on a new linetype when changed in the "Layer Properties Manager". This is an example of a parent control. If the object's linetype property is changed with a child override, then the linetype cannot be altered at the parent level.

  Similarly, users have the ability to apply child overrides to individual labels. This can be accomplished by selecting a label(s), right clicking and selecting "Edit Label Properties". Additional child overrides may be applied to curve labels using the "CAD Properties" option when right clicking curve labels.

  Two popular parent level utilities, available when right clicking labels, are the ability to edit the objects label style and set that particular label style current.

- Run through the same process for the curve label styles and curve labeling so that your drawing matches the reference figure at the end of this chapter.

- Set up a dimension style for L80-060X and while on the "C-ANNO-DIMS" layer dimension the drawing in the same fashion that the drawing is dimensioned for this assignment.

- Pick the layout tab currently titled "Layout1" so that this layout becomes active. Right click on this tab and choose to rename the layout tab to "Plat". Now that you are in paperspace, change to the layer "C-ANNO-VPRT" and create a landscape viewport 11'×8.5' at a 60 scale. Your drawing will not entirely fit in this viewport at this scale. The orientation must be altered. We will address this in a future project with the implementation of a "Dview Twist". In the mean time, center it as best you can.

- Save and exit this drawing. You have successfully completed this assignment.

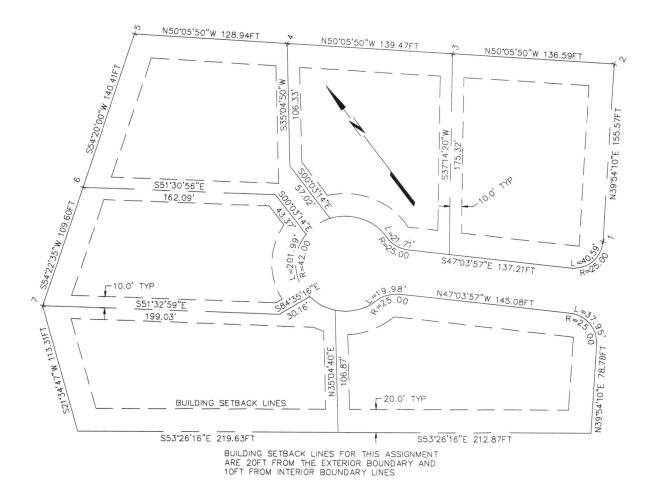

Reference Figure: Assignment 8

NOTES:

Assignment #9
Line and Curve Tables

Recommended Assignments Prior to Working this Assignment:

Assignments 1-8

Required Assignments Prior to Working this Assignment:

Assignment 7

Goals and Objectives

      Land Desktop line labeling can be such a convenient means of displaying object information that users sometimes tend to over label. Remember that drawings are created to convey information to others and it is not desirable to create a drawing which is too difficult to decipher. In Assignment #8, we chose not to label the building setback lines. Instead, a few of them were dimensioned and noted as typical. This helps generate a clear, concise drawing. Another way that users can make their drawings more readable is through tabling line and curve data using Land Desktop. We will take advantage of these easy to use routines in this assignment.

Exercise Instructions

- Logon to your workstation and begin a session of Land Desktop. If Land Desktop has been configured on your workstation to display the "Start Up" dialogue box, then cancel this feature so that the AutoCAD model space environment is displayed. If launching Land Desktop brings you directly into the AutoCAD model space environment, then Land Desktop has been configured to begin without the startup dialogue. If the AutoCAD Map "Task Pane" is shown, then close this dialogue box as well. Users that have the ability to customize their profiles can turn these features off permanently by following the procedure described in Assignment #1.

- Begin a new drawing by selecting "File→New". Choose to create a new project by picking the button labeled "Create Project". Select the "Prototype:" "Default (Feet)" from the drop down list. Give the project the "Name:" "Assign09-##" where the "##" reflects the number which was assigned to you on the first day of class. If the number assigned to you is a single digit (for example 4 as opposed to 14), then enter a zero preceding the single digit (such as 04). Write "Line and Curve Tables" in the "Description:" and "Keywords:" areas. This will serve as a project summary which may be used to find or filter projects using the Project Manager.

- Select "OK" to bring you back to the "New Drawing" dialogue box and type the "Name:" "Assign09-##" where the "##" reflects the number which was assigned to you on the first day of class. Make sure that the "Project Name:" displays the correct project.

- Select the "Acad.dwt" template and then "OK" to generate a new drawing in the "DWG" folder of the current project. After selecting the "OK" button, you may be prompted to save changes to the previous drawing session. Since there were no objects in the previous drawing session worth saving, choose not to save changes.

- You will be prompted with the "Load Settings" dialogue box which displays a list of drawing setups to choose from. These setups are saved back to the path determined by the Network Administrator who installed the software. The default path is to the local machine and is displayed in the dialogue box. We will set up our parameters manually, so select "Next" to set up your project with the following parameters:

    <u>Units Area</u>
    Linear Units = Feet
    Angle Units = Degrees
    Angle Display Style = Bearings
    Display Precision Linear = 2
    Display Precision Elevation = 2
    Display Precision Coordinate = 5
    Display Precision Angular = 4
    "Next"

    <u>Scale Area</u>
    Horizontal = 60
    Vertical = 1
    Paper Size = 8 x 11(A)
    "Next"

    <u>Zone Area</u>
    "Next"

    <u>Orientation Area</u>
    "Next"

    <u>Text Style Area</u>
    Leroy.stp
    L80

- Then select "Finish" and a screen will display providing the user with a summary of the settings chosen. Review the settings and select "OK".

Although users have the ability to save settings for retrieval with future projects, assignments in this text require that you set up the parameters manually for practice in each project.

- You will be prompted with the "Create Point Database" dialogue box. Accept the default settings by choosing "OK".

- Create several new layers with the following parameters:

Layer	Color	Linetype
C-ANNO-TABL	Red	Continuous
C-ANNO-TEXT	White	Continuous
C-ANNO-VPRT	White	Continuous

- Insert the drawing Assign07-## at the coordinates 0,0 at a scale of 1 and with a rotation angle of 0. Zoom extents and explode the drawing one time.

- Select "Labels→Edit tag styles". Set up Line and Curve tag styles named "Tag Labels Above" and "Tag Labels Below" to yield the same labeling results that are shown in the figure at the end of this chapter. (The "Edit tag styles" routine works similar to the "Edit label styles" routine, so if you are having difficulty setting up your tag style, refer to Assignment #8.) An example of the "Tag Labels Above" style is shown below for reference.

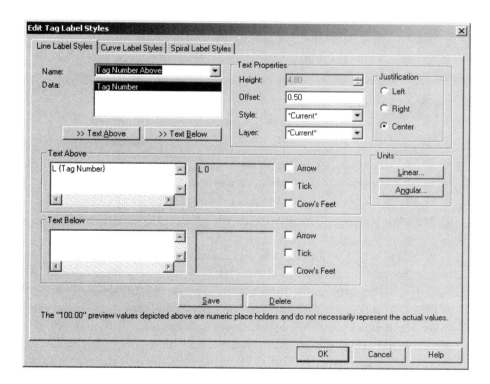

- In the menu "Labels→Settings→Line Labels" set the appropriate tag label style current.

Introduction to Land Desktop 2007

- Set the current AutoCAD text style to an L80 set up for a 60 scale and set the "C-ANNO-TEXT" layer current. Proceed to label the boundary lines in the same fashion that the boundary is labeled for this assignment using the command "Labels→Add Tag Labels". The linework can also be labeled with the current tag style by selecting the line, right clicking and choosing the option "Add Tag Label".

- Tabling is one of the features which changed in a previous release of the software (Land Desktop 2005). The new software still supports the older table format for those users who have set up table templates having a particular look, but the software has been improved to also include the ability to table information with AutoCAD's tabling feature. Select "Labels→Add Tables→Line Table" and fill out the table to reflect the following:

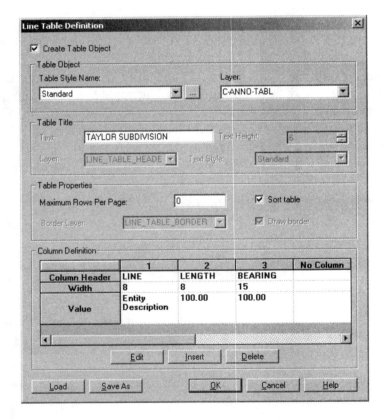

The "Create Table Object" check box instructs Land Desktop to create tables using the tabling feature in AutoCAD. If the check box is **NOT** toggled on, tabling will be created as it has been in previous releases of the software. Whereas the older tabling feature saves tables back to the path "C:\Program Files\Land Desktop 2007\Data\labels", AutoCAD 2007 Tables are saved with the drawing. Therefore, users that wish to have a consistent look among tables using the AutoCAD 2007 tabling feature will need to either use AutoCAD's Design Center routine to read table formats into their drawings, or create prototype (Template) drawings that have the table formats built in to them.

- While in the "Line Table Definition" dialogue box, select the button to the right of the word "Standard" to launch the "Table Style" manager shown below:

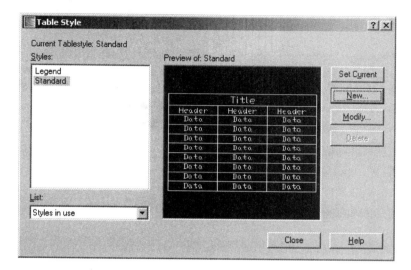

- Create a new table style by selecting the "New" button on the right side of the table style manager. Name the table style "Tag Table" and choose to start with the default table, which will likely be named "Standard". Select the "Continue" button to proceed and set up the table as follows:

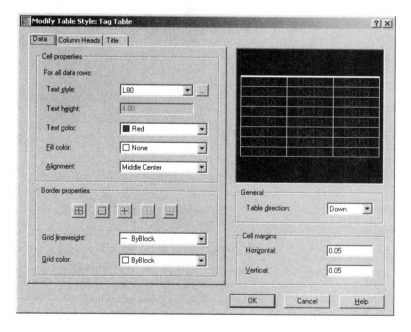

Introduction to Land Desktop 2007

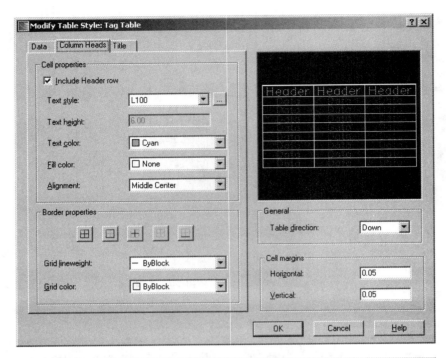

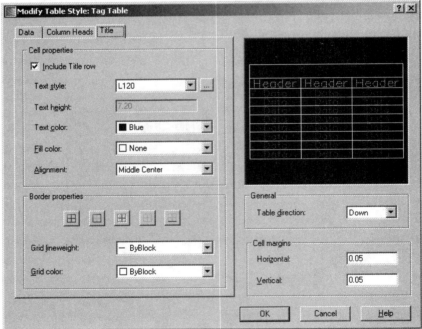

After creating your new table style, select "OK" to bring you back to the Table Style Manager, "Set Current", and then "Close" to bring you back to the "Line Table Definition" dialogue box.

- Edit the characteristics for the Bearing column by selecting it and choosing the "Edit" option below and to the left of the Bearing column. A graphic displaying this dialogue box is shown on the following page for reference.

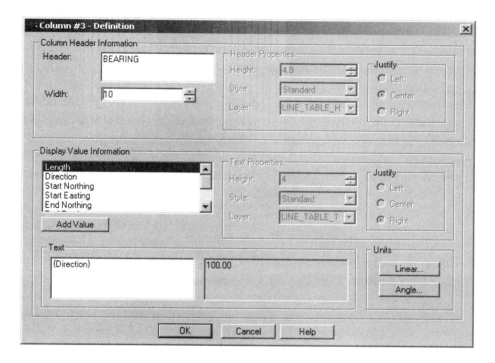

- Select "OK" to bring you back to the "Line Table Definition" dialogue box and "OK" again to insert your table into the model space environment at the coordinate pair (1820600,290500). If you need to make adjustments to your table, then you may do so using the command "Labels→Edit Tables→Edit Table Layout" and selecting any object affiliated with the table. When users elect to make changes to their table, the dialogue box header reads "Modify Table Style" as opposed to "New Table Style". Edits to tables created using AutoCAD 2007's tabling feature as were created in this assignment will update automatically. If tables are created without using the new tabling feature, they will need to be reinserted after making edits using the command "Labels→Edit Tables→Re-Draw Table". This command is also required to update table data after altering line work having tag labels.

- Now construct a curve table similar to the way that you created the line table. Notice as you enter the "Curve Table Definition" dialogue box that the "Tag Table" style created with the line table is present in the drop down dialogue box. Choose this table style and insert your curve table at the coordinate pair (1820900,290500). You may need to increase the width of your curve, radius and length columns so that your subdivision title reads on a single line. This can be done while inside the "Curve Table Definition" dialogue box, or by simply using grips to widen the table after insertion.

- Pick the layout tab currently titled "Layout1" so that this layout becomes active. Right click on this tab and choose to rename the layout tab to "Plat". Now that you are in paperspace, change to the layer "C-ANNO-VPRT" and create 3 viewports that fit on an 8 ½ inch x 11 inch sheet of paper which display the map and two tables.

- Size the upper viewport at 11' x 6.5' and display the plan view. Since the map doesn't quite fit in this viewport at the desired scale, we are going to visually twist the drawing through the viewport. While on the layout tab, set your environment to "Model Space" so that the cursor is inside the 11' x 6.5' viewport. At the command line type "Dview". When prompted to select entities, type "All" (you don't need to type "All" to have all objects visually twisted; the objects selected at this prompt show up in the dview twist preview) and enter twice. When prompted with the next menu, type "TW" for twist and then enter again. Enter an angle of 36 degrees and press the "Enter" key twice to exit the command.

- Since your tables are in different viewports, they will remain orthogonal in your plan view layout. Make sure that your viewport scale is 60. Your global linetype scale for paperspace plotting should be set to "1". The AutoCAD system variable "PSLTSCALE" should also be set to "1".

- Save and exit this drawing. You have successfully completed this assignment.

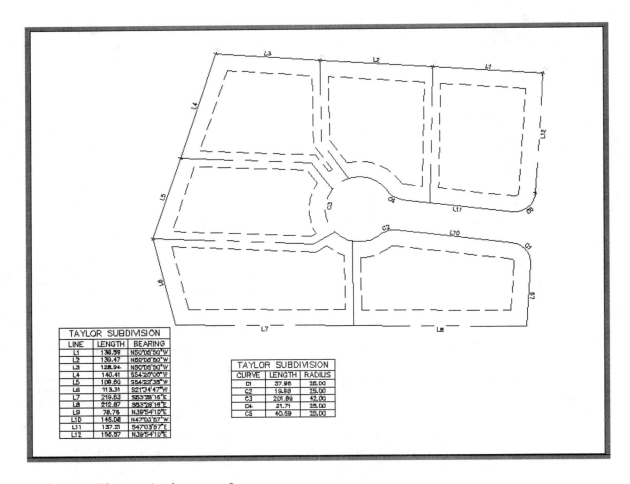

Reference Figure: Assignment 9

> Assignment #10
>
> ## *Point Labeling with Description Keys*

Recommended Assignments Prior to Working this Assignment:

Assignments 1-9

Required Assignments Prior to Working this Assignment:

None

Goals and Objectives

      When AEC points are inserted into the drawing environment, they display the point marker, number, elevation and raw description by default. The point settings can be configured to show any combination of number, elevation and raw description or configured not to show any of them at all. The "Point Settings" configuration only affects points about to be inserted into a drawing environment. If it is desirable to change the appearance of points already in the drawing environment without reinserting them, then users must use the command "Points→Edit Points→Display Properties".

      Just as we labeled lines in previous lessons, points can also be labeled. One advantage to using point labels as opposed to standard point markers (Marker, number, elevation, Description) is that users can display more information with point labels than a standard point marker. Point labels can be used to label points with any text string, and/or a variety of other information such as Northings and Eastings which may be extracted from the external COGO (Coordinate Geometry) points database. Point labeling also gives users the flexibility of including a symbol in the form of a block with their label.

      One popular point labeling application occurs with tree labeling. A label style can be set up so that points shot on trees only display a tree number, elevation and symbol. If tree points are inserted into the drawing environment using point labels in the above referenced application, then the tree block comes in automatically along with the tree number and elevation.

Introduction to Land Desktop 2007

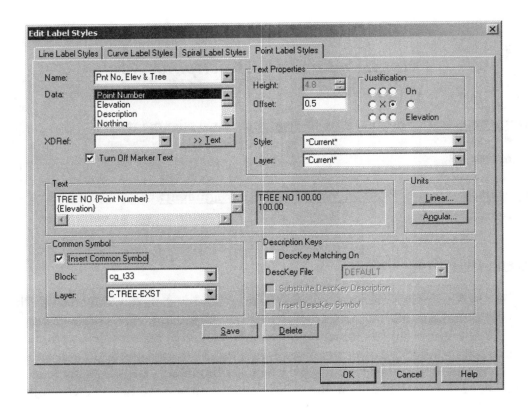

The point label style shown above will generate a point label displaying the tree number, elevation and COGO tree symbol t33 upon point(s) insertions into the drawing environment. The symbol block path is specified on the "Insert" tab of the "Point Settings" dialogue box. Point labeling must also be enabled on the "Insert" tab if it is desired that Land Desktop label points.

Users might also choose to set up additional label styles for importing utility points. As an example, a label style could be set up for inserting points and symbols locating fire hydrants. In this case, a user may only want to display the fire hydrant and spot elevation.

Although point labels in the above applications will save users time, it is not desirable to set different label styles current as points are inserted to the drawing environment by group or description. In the above referenced examples, users would have needed to set each of the label styles current prior to bringing in points respective to the point labels. The tree label style would have needed to be set current prior to inserting tree points, just as the fire hydrant label style would have needed to be set current prior to inserting fire hydrant points.

Before you set up point label styles for every manhole, mailbox, catch basin, tree, sign and bush, there is another Land Desktop feature that enables users to automate labeling to some degree. This feature is referred to as "Description Keys". "Description Keys" are used in conjunction with point labeling and are quite powerful. Whereas point labeling allows users to label batches of points inserted to the drawing environment having common parameters, the implementation of "Description Keys" permits users to read in and label all points at once regardless of what they are. "Description Keys" permit block symbol & layer mapping as well as scaling. Symbol blocks can be scaled by parameters in the point descriptor. For instance, the descriptor "T O 7" can be read in and manipulated using "Description Keys" and point labeling to read "7 inch Oak" in the drawing, be assigned a tree number, be mapped to a tree symbol and have the symbol scaled by a factor of 7 upon its insertion into the drawing environment. In this example, the layer containing the point marker could be placed on one layer, while the point label be placed on a second layer and the block symbol placed on a third layer. In the above referenced example, the description "T O 7" represents the "Raw Description" and "7 inch Oak" takes the place of the "Full Description".

In this assignment, you will create a point label style for trees, and a point label style for all other points to be inserted into the drawing environment. You will also create a description key file to map descriptors with the label styles, layers, text and symbol blocks.

Exercise Instructions

- Logon to your workstation and begin a session of Land Desktop. If Land Desktop has been configured on your workstation to display the "Start Up" dialogue box, then cancel this feature so that the AutoCAD model space environment is displayed. If launching Land Desktop brings you directly into the AutoCAD model space environment, then Land Desktop has been configured to begin without the startup dialogue. If the AutoCAD Map "Task Pane" is shown, then close this dialogue box as well. Users that have the ability to customize their profiles can turn these features off permanently by following the procedure described in Assignment #1.

- Begin a new drawing by selecting "File→New". Choose to create a new project by picking the button labeled "Create Project". Select the "Prototype:" "Default (Feet)" from the drop down list. Give the project the "Name:" "Assign10-##" where the "##" reflects the number which was assigned to you on the first day of class. If the number assigned to you is a single digit (for example 4 as opposed to 14), then enter a zero preceding the single digit (such as 04). Write "Point Labeling with Description keys" in the "Description:" and "Keywords:" areas. This will serve as a project summary which may be used to find or filter projects using the Project Manager.

- Select "OK" to bring you back to the "New Drawing" dialogue box and type the "Name:" "Assign10-##" where the "##" reflects the number which was assigned to you on the first day of class. Make sure that the "Project Name:" displays the correct project.

- Select the "Acad.dwt" template and then "OK" to generate a new drawing in the "DWG" folder of the current project. After selecting the "OK" button, you may be prompted to save changes to the previous drawing session. Since there were no objects in the previous drawing session worth saving, choose not to save changes.

- You will be prompted with the "Load Settings" dialogue box which displays a list of drawing setups to choose from. These setups are saved back to the path determined by the Network Administrator who installed the software. The default path is to the local machine and is displayed in the dialogue box. We will set up our parameters manually, so select "Next" to set up your project with the following parameters:

    Units Area
       Linear Units = Feet
       Angle Units = Degrees
       Angle Display Style = Bearings
       Display Precision Linear = 2
       Display Precision Elevation = 2
       Display Precision Coordinate = 5
       Display Precision Angular = 4
       "Next"

    Scale Area
       Horizontal = 60
       Vertical = 1
       Paper Size = 8 x 11(A)
       "Next"

    Zone Area
       "Next"

    Orientation Area
       "Next"

    Text Style Area
       Leroy.stp
       L80

- Then select "Finish" and a screen will display providing the user with a summary of the settings chosen. Review the settings and select "OK". Although users have the ability to save settings for retrieval with future projects, assignments in this text require that you set up the parameters manually for practice in each project.

- You will be prompted with the "Create Point Database" dialogue box. Accept the default settings by choosing "OK".

- Create several new layers with the following parameters:

Layer	Color	Linetype
C-ANNO-VPRT	White	Continuous
C-PNTS-TREE	Red	Continuous
C-PNTS-UTIL	Green	Continuous
C-TREE-EXST	Green	Continuous
C-UTIL-SYMB	Cyan	Continuous

- Set your point settings (Refer to Assignment 3 if you can't remember how this was accomplished) so that your point text comes in with the L80 text style. Using the "+" marker, your marker height should be set at ½ the model space text height. The corresponding marker height for a text height of 4.8 units is 2.4. Set a text rotation angle of 30 degrees.

- Set your point settings to allow the use of point labels by selecting "Points→Point Settings→Insert (Tab) " and turning on the checkbox to "Use Current Point Label Style When Inserting Points". The "Search Path for Symbol Block Drawing Files" is the path that Land Desktop uses to find blocks which are inserted with point labeling. Change the path to reflect the COGO folder for the Land Desktop 2007 symbol manager folder. Although the location of this folder can be altered with the software installation the default installation path is:

  "c:\documents and settings\all users\application data\autodesk\autodesk land desktop 2007\r16.2\data\symbol manager\cogo\"

  Refer to the graphic below for reference.

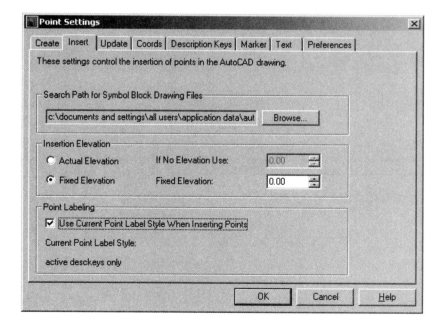

- Launch the description key manager through the menu "Points→Point Management→Description Key Manager". After the dialogue box opens, select "Manager→Create Desckey File" and give it the file name "Assign10". Right select on the newly created description key file and select "Create Desckey". Proceed to fill out information as shown in the dialogue box below. In the "Desckey Code" area type "t*". Fill in areas as shown below for three different Description keys. Keep in mind that description keys are case sensitive.

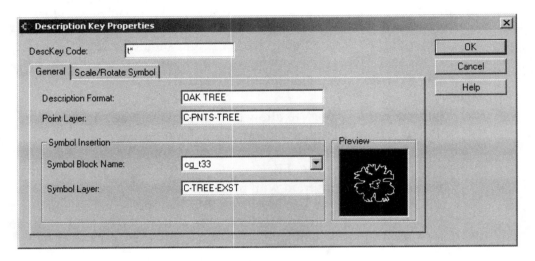

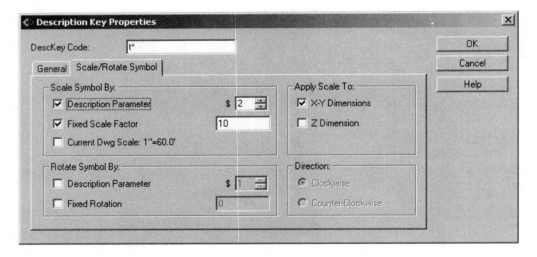

Point Labeling with Description Keys

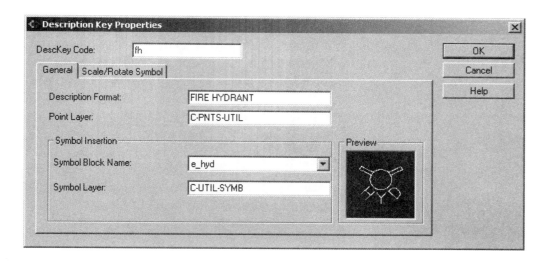

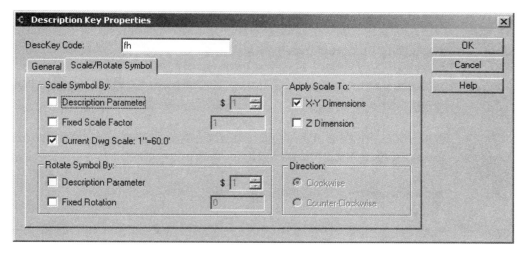

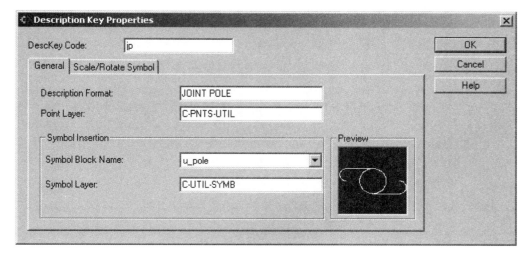

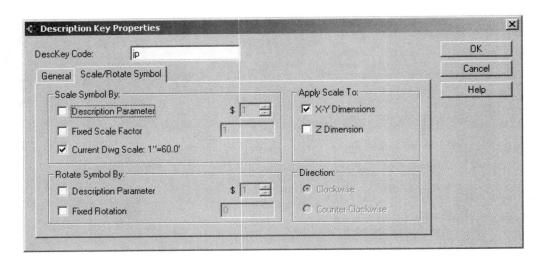

- The "Description Parameter", "2", found on the "Scale/Rotate Symbol", tab instructs Land Desktop to read the second parameter past the base descriptor for scaling the associated block. In the description "Tree O 7", the base descriptor is "Tree". The first parameter is "O" for Oak tree. The second parameter is the number "7". The $2 instructs Land Desktop to scale the symbol by a factor of 7.

- When you are finished entering description key information, the dialogue box should look something like the one below.

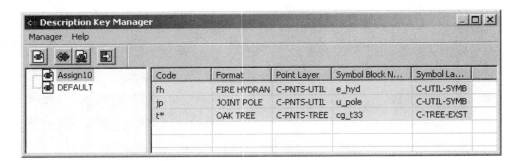

- Exit the description key manager and select "Labels→Edit Label Styles→Point Label Styles" to create two new point label styles as shown below named "Elevation, Block" and "Point, Elevation, Block".

Point Labeling with Description Keys

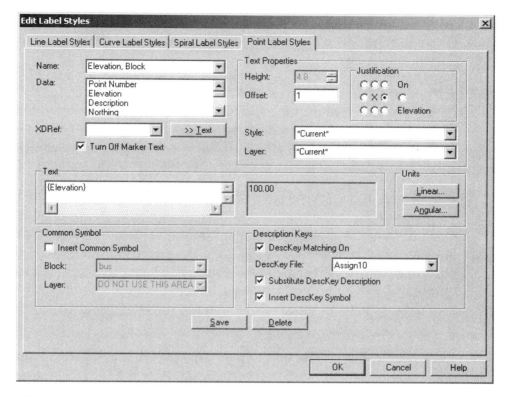

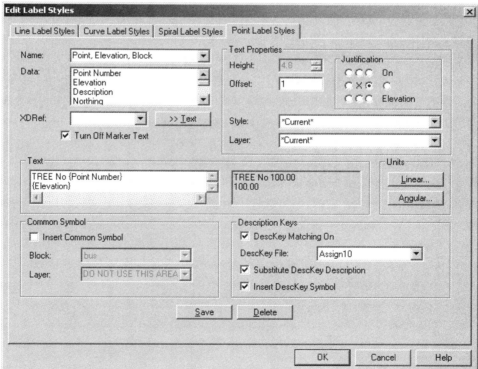

"DescKey Matching On" must be checked to allow the point label style to obtain information from the description key file. Specify the "Desckey File: " which you just created for this assignment.

10.9

A "Common Symbol" is not required, because the symbol will be inserted to the CAD environment using the Description Key file. In order to accomplish this, make sure the check box "Substitute DescKey Symbol" is checked.

The "Substitute DescKey Description" checkbox instructs Land Desktop to alter the Point(s) full description to take on the description specified in the "Description Format" area for each Description Key in the Description Key Manager.

- Select "Labels→Settings→Point Labels" and set the "Elevation, Block" label style which you just created current. Specify a rotation angle of 30 degrees.

- Set the "C-PNTS-UTIL" layer current and select "Points→Import/Export Points→Import Points" to read in your project points which correspond to this lesson using the "PNEZD" comma delimited format. Lesson files may be downloaded from www.schroff1.com. As you read these points in, add them to a new point group named "All_Points". Choose to overwrite points when prompted with the "Cogo Database Import Options". Zoom extents, and you should see spot elevations and symbols in the drawing environment.

- In "Labels→Settings→Point Labels" set the "Point, Elevation, Block" label style current. While on the "C-PNTS-TREE" layer, insert only points having the description tree. Refer to Assignments 4 & 5 if you can't remember how this is done. Choose to replace all points when prompted. The tree spot elevations are now on their respective layer. Tree symbols have been inserted into the drawing environment and scaled.

- Pick the layout tab currently titled "Layout1" so that this layout becomes active. Right click on this tab and choose to rename the layout tab to "Plat". Now that you are in paperspace, change to the layer "C-ANNO-VPRT" and create a landscape viewport 11′×8.5′ at a 60 scale for the map.

- Save and exit this drawing. You have successfully completed this assignment.

Point Labeling with Description Keys

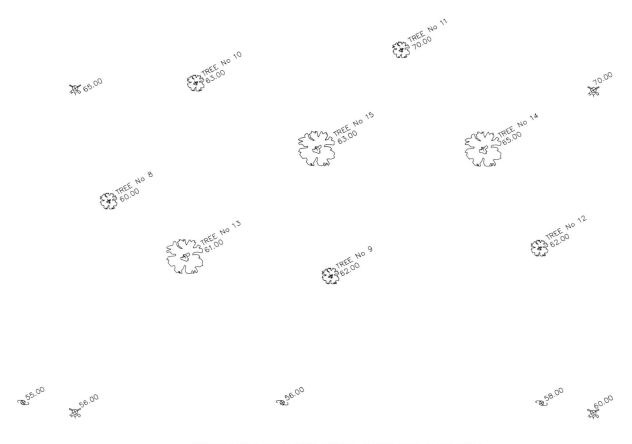

Reference Figure: Assignment 10

10.11

NOTES:

Assignment #11

Parcel Computations

Recommended Assignments Prior to Working this Assignment:

Assignments 1-10

Required Assignments Prior to Working this Assignment:

None

Goals and Objectives

Suppose that an AutoCAD user wanted to compute the area of a figure bounded by several lines. This could be accomplished using the AutoCAD area command. If the figure contained an arc, the user might construct a polyline of the bounding entities and list the polyline to obtain an area. The figure's area could also be listed using the Land Desktop commands "Inquiry→Area By Lines/Curves", "Inquiry→Area By Polylines" or "Inquiry→Area By Points".

Consider for a moment how figures are drawn in the AutoCAD environment. If an existing 4 sided parcel is mapped by drawing 4 lines between 4 found pipes which demark the parcel's property corners, then this will undoubtedly generate a closed figure. If the bearings and distances for each course are computed and written down, the figure could be reconstructed from the written bearings and distances. However, the geometry of the figure, to some degree, will be dictated by the precision of the written bearings and distances. If distances are recorded to the nearest foot, then the reconstructed figure's geometry will be different than a figure reconstructed from distances recorded to the nearest hundredth of a foot. In addition, if a non rectangular figure having distances rounded to the nearest foot is reconstructed, there is a low probability that the figure will close.

The Land Desktop "Parcel" routine computes areas bounded by points, lines/arcs or polylines. If the existing drawing linework does not close within a preset tolerance (approximately .000001ft), the user is issued an error message and given the option to close the figure by adding an additional segment (Force Closure) or canceling the command to adjust the figure.

The routine runs an inverse (Computation of bearings and distances based on coordinates) along rounded bearings and distances of the existing linework where the rounded values are established by the precision settings in the drawing setup. The parcel routine gives users the ability to label parcels and prepare report summaries. Users can report "Areas" based on the drawing linework, and "Error Closure" based on the rounded bearings and distances established by the Land Desktop precision settings. In this assignment users will be introduced to parcel computations.

Exercise Instructions

- Logon to your workstation and begin a session of Land Desktop. If Land Desktop has been configured on your workstation to display the "Start Up" dialogue box, then cancel this feature so that the AutoCAD model space environment is displayed. If launching Land Desktop brings you directly into the AutoCAD model space environment, then Land Desktop has been configured to begin without the startup dialogue. If the AutoCAD Map "Task Pane" is shown, then close this dialogue box as well. Users that have the ability to customize their profiles can turn these features off permanently by following the procedure described in Assignment #1.

- Begin a new drawing by selecting "File→New". Choose to create a new project by picking the button labeled "Create Project". Select the "Prototype:" "Default (Feet)" from the drop down list. Give the project the "Name: "Assign11-##" where the "##" reflects the number which was assigned to you on the first day of class. If the number assigned to you is a single digit (for example 4 as opposed to 14), then enter a zero preceding the single digit (such as 04). Write "Parcel Computations" in the "Description:" and "Keywords:" areas. This will serve as a project summary which may be used to find or filter projects using the Project Manager.

- Select "OK" to bring you back to the "New Drawing" dialogue box and type the "Name:" "Assign11-##" where the "##" reflects the number which was assigned to you on the first day of class Make sure that the "Project Name:" displays the correct project.

- Select the "Acad.dwt" template and then "OK" to generate a new drawing in the "DWG" folder of the current project. After selecting the "OK" button, you may be prompted to save changes to the previous drawing session. Since there were no objects in the previous drawing session worth saving, choose not to save changes.

- You will be prompted with the "Load Settings" dialogue box which displays a list of drawing setups to choose from. These setups are saved back to the path determined by the Network Administrator who installed the software. The default path is to the local machine and is displayed in the dialogue box. We will set up our parameters manually, so select "Next" to set up your project with the following parameters:

Units Area
> Linear Units = Feet
> Angle Units = Degrees
> Angle Display Style = Bearings
> Display Precision Linear = 2
> Display Precision Elevation = 2
> Display Precision Coordinate = 5
> Display Precision Angular = 4
> "Next"

Scale Area
> Horizontal = 60
> Vertical = 1
> Paper Size = 8 x 11(A)
> "Next"

Zone Area
> "Next"

Orientation Area
> "Next"

Text Style Area
> Leroy.stp
> L80

- Then select "Finish" and a screen will display providing the user with a summary of the settings chosen. Review the settings and select "OK". Although users have the ability to save settings for retrieval with future projects, assignments in this text require that you set up the parameters manually for practice in each project.

- You will be prompted with the "Create Point Database" dialogue box. Accept the default settings by choosing "OK".

- Create several new layers with the following parameters:

Layer	Color	Linetype
C-ANNO-LOTS	White	Continuous
C-ANNO-VPRT	White	Continuous
C-PROP-BNDY	Magenta	Continuous
C-PROP-LOTS	Cyan	Continuous
C-PROP-LOTS-PLIN	Blue	Continuous
C-PNTS-TOPO	Green	Continuous

- Set your point settings (Refer to Assignment 3 if you can't remember how this was accomplished) so that your point text comes in with the L80 text style.

Using the "+" marker, your marker height should be set at ½ the model space text height. The corresponding marker height for a text height of 4.8units is 2.4. Set a text rotation angle of 30 degrees.

- Set your "C-PNTS-TOPO" layer current and read in your project points from the comma delimited file which corresponds to this lesson. Lesson files may be downloaded from http://www.schroff1.com. As you read these points in, add them to a new point group named "All_Points". Refer back to previous assignments if you can't remember how this was accomplished. These points represent the vertices of the property that you are going to map.

- Set up the "Parcels→Parcel Settings" dialogue box to reflect the following:

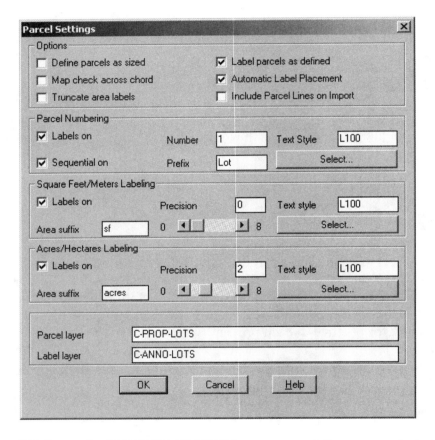

- While on the "C-PROP-BNDY" layer connect the survey shots as shown in the figure at the end of this chapter. You should be able to draw the boundary information using AutoCAD lines and arcs.

- Connect the points making up the subdivision boundary on the layer "C-PROP-BNDY". Interior lot lines should be drawn on the layer "C-PROP-LOTS".

- Select "Parcels→Define from Lines/Curves" to label the first 2 lots as shown in the figure at the end of this chapter. When the command is invoked, you will be prompted to select an entity. Although the entity selected must be either a line or an arc, it doesn't matter which object is selected for the purpose of creating a parcel. However, when selecting the first entity, users should consider that the first entity selected will be the first entity that appears when reporting parcel related information using the parcel manager. Select one of the objects making up "Lot 1". When prompted to "Select Objects", select all objects (Lines and Arcs) that make up "Lot 1" and press "Enter". If your linework was drawn appropriately, then the parcel will have been written to the external database and labeled in the model space environment. One additional "Enter" will end the command. Define Lot 2 using the same procedure that you used to define Lot 1.

- Select "Parcels→Define from Points" to label Lot 3 as indicated in the figure at the end of this chapter. Keep in mind that when Land Desktop prompts a user for "Area first point", Land Desktop is requesting that users select a lot corner using AutoCAD object snaps. The prompt can be changed to point number by using the toggle ".P". This lot is made up of points 6,10,11,12 and 7.

- Select "Parcels→Define from Lines/Curves" to label lots 4 and 5 as shown in the figure at the end of this chapter.

- Copy the lines bounding lot 6 to the layer "C-PROP-LOTS-PLIN". Use the "Pedit" command with the "Join" option to create a polyline from the copied lines and define this parcel using the command "Parcels→Define from Polylines".

- Select "Parcels→Parcel Manager→Select All→Map Check→Print to File" and save as "Mapcheck.txt" in the "Lots" directory for the corresponding project. Launch a session of notepad and view the file to assure that it is present and correct. Notice that the closure error is displayed at the bottom of each lot calculation.

- Pick the layout tab currently titled "Layout1" so that this layout becomes active. Right click on this tab and choose to rename the layout tab to "Plat". Now that you are in paperspace, change to the layer "C-ANNO-VPRT" and create a landscape viewport 11'×8.5' at a 60 scale. Your drawing will not entirely fit in this viewport at this scale. The orientation must be altered.

- In order to get the map to fit, we are going to visually twist the drawing while in the viewport. While on the layout tab, set your environment to "Model Space" so that you are able to see your floating viewport and your map. At the command line type "Dview". When prompted to select entities, type "All" and enter twice. When prompted with the next menu, type "TW" for twist and then enter again. Enter an angle of 15 degrees and hit the enter key to exit the command.

- Make sure that your viewport scale is 60. Your global linetype scale for paperspace plotting should be set to "1". The AutoCAD system variable "PSLTSCALE" should also be set to "1".

- Save and Exit this drawing. You have successfully completed this assignment.

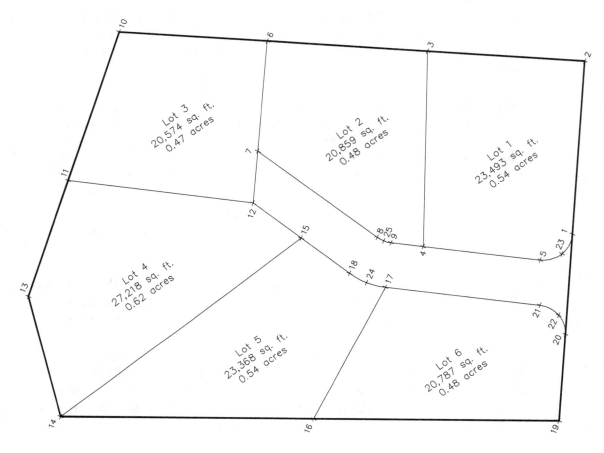

Reference Figure: Assignment 11

Assignment #12
Creating a Surface

Recommended Assignments Prior to Working this Assignment:

Assignments 1-7

Required Assignments Prior to Working this Assignment:

None

Goals and Objectives

In the past several assignments, we learned how to import points to a project and create a 2-dimensional topographic map representing our surveyed area. In this assignment we will get a first glance at adding the 3rd dimension to our data. The 3rd dimension gives us the ability to create cross sections and profiles. It also gives us the ability to create contours which are a 2 dimensional representation of a 3 dimensional surface.

We will direct Land Desktop to create a 3 dimensional surface from our project point data. In older versions of the Civil/Survey software, the surface was called a TIN. The **Surface** is a network of irregularly shaped triangles where, in this assignment, the project points make up the triangles vertices. There are actually several data types which can be used to generate a surface, some of which will be explored in future assignments. In this assignment we will focus on creating a surface from project points.

Upon examining some earthwork volumes generated by your supervisor, you realize that the volumes are not rational. When you confront your supervisor and inform her of your familiarity of the project and let her know that the volumes that she has come up with are not reasonable, given the site geometry, she assures you that she is performing the computations correctly. You suspect that one of the surfaces which she is computing the volumes from contains erroneous data. In an effort to resolve the problem you rebuild the surface from the data which your supervisor was given in an effort to identify the problem.

Exercise Instructions

- Logon to your workstation and begin a session of Land Desktop. If Land Desktop has been configured on your workstation to display the "Start Up" dialogue box, then cancel this feature so that the AutoCAD model space environment is displayed. If launching Land Desktop brings you directly into the AutoCAD model space environment, then Land Desktop has been configured to begin without the startup dialogue. If the AutoCAD Map "Task Pane" is shown, then close this dialogue box as well. Users that have the ability to customize their profiles can turn these features off permanently by following the procedure described in Assignment #1.

- Begin a new drawing by selecting "File→New". Choose to create a new project by picking the button labeled "Create Project". Select the "Prototype:" "Default (Feet)" from the drop down list. Give the project the "Name:" "Assign12-##" where the "##" reflects the number which was assigned to you on the first day of class. If the number assigned to you is a single digit (for example 4 as opposed to 14), then enter a zero preceding the single digit (such as 04). Write "Creating a Surface" in the "Description:" and "Keywords:" areas. This will serve as a project summary which may be used to find or filter projects using the Project Manager.

- Select "OK" to bring you back to the "New Drawing" dialogue box and type the "Name:" "Assign12-##" where the "##" reflects the number which was assigned to you on the first day of class. Make sure that the "Project Name:" displays the correct project.

- Select the "Acad.dwt" template and then "OK" to generate a new drawing in the "DWG" folder of the current project. After selecting the "OK" button, you may be prompted to save changes to the previous drawing session. Since there were no objects in the previous drawing session worth saving, choose not to save changes.

- You will be prompted with the "Load Settings" dialogue box which displays a list of drawing setups to choose from. These setups are saved back to the path determined by the Network Administrator who installed the software. The default path is to the local machine and is displayed in the dialogue box. We will set up our parameters manually, so select "Next" to set up your project with the following parameters:

    <u>Units Area</u>
       Linear Units = Feet
       Angle Units = Degrees
       Angle Display Style = Bearings
       Display Precision Linear = 2
       Display Precision Elevation = 2
       Display Precision Coordinate = 5
       Display Precision Angular = 4
       "Next"

    <u>Scale Area</u>
       Horizontal = 40
       Vertical = 10
       Paper Size = 24 x 36(D)
       "Next"

<u>Zone Area</u>
"Next"

<u>Orientation Area</u>
"Next"

<u>Text Style Area</u>
Leroy.stp
L80

- Then select "Finish" and a screen will display providing the user with a summary of the settings chosen. Review the settings and select "OK". Although users have the ability to save settings for retrieval with future projects, assignments in this text require that you set up the parameters manually for practice in each project.

- You will be prompted with the "Create Point Database" dialogue box. Accept the default settings by choosing "OK".

- Create several new layers with the following parameters:

<u>Layer</u>	<u>Color</u>	<u>Linetype</u>
C-ANNO-VPRT	White	Continuous
C-CNST-LINE	Green	Continuous
C-PNTS-TOPO	Red	Continuous

- Set your point settings so that your point text comes in with the L80 text style. Using the "+" marker, your marker height should be set at ½ the model space text height. The corresponding marker height for a text height of 3.2 units is 1.6. Set a text rotation angle of 30 degrees.

- Set the "C-PNTS-TOPO" layer current and select "Points→Import/Export Points→Import Points" to read in the comma delimited point file from the file which corresponds to this lesson. Lesson files may be downloaded from www.schroff1.com. As you read these points in, add them to a new point group named "All_Points". Refer back to previous assignments if you can't remember how this was accomplished. Now zoom extents to observe these points. These points represent the vertices of the new surface that you are about to create.

- Launch the terrain model explorer by selecting "Terrain→Terrain Model Explorer". The terrain model explorer is a manager that keeps track of the project surfaces.

- Create a new surface by selecting the menu "Manager→Create Surface" or using your mouse and right clicking on the manila folder labeled "Terrain" and electing to create a new surface. A new surface should have been created as a file (Surface1) in the terrain folder. To view the file, click the "+" symbol to the left of the Terrain folder. Right click on the surface name and choose to "Rename" it to "EG". Pick on the renamed surface with the Right mouse button and select "Properties" to view the properties of the surface. Type in a Description "Existing Ground from project points". Select "OK" to exit the properties dialog box.

- We have now created the name and file for the surface. We have to choose the data to be used for surface generation. There are many types of data that can be used to build a surface. In this assignment, we will focus on building a surface from project point data. In order to do this click on the "+" symbol to the left of the surface name to view the surface data types. Right click on "Point Groups" to add the point group "All_points" that you created when you imported points into this project. This will allow Land Desktop to create a surface from the data in this point group.

- In order for surfaces to be built using point groups as a data type, point groups must have updated status. To update the status of point groups users need to launch the point group manager "Points→Point Management→Point Group Manager" and update point groups being used as a data type for surface generation. Use the menu "Manager→Update All Point Groups" to accomplish this.

- Now that we have told Land Desktop what to use for building the surface, we can tell it to construct the surface. Launch the "Terrain Model Explorer", right click on the surface name, "EG", and Select the "Build" option. You will be shown a dialogue box which summarizes the settings. Accept the settings by choosing "OK" and Land Desktop will build the surface.

- In order to view the surface which you have just built, while in the "Terrain Model Explorer", right click on the surface name, "EG", and choose the menu "Surface Display→Quick View". Minimize or close the terrain model explorer and notice the white vectors on your screen. These vectors are not entities in your drawing. They are a visual representation of your surface. Any AutoCAD command which results in screen regeneration will make these vectors disappear. Type "R" for redraw and see for yourself.

- Lets have a look at the surface in three dimensions. Type "VP" at the command line and enter an angle of 275 degrees from the x-axis and 30 degrees from the xy- plane. Then select "OK". Zoom to a scale of "1/10".

- To see the vector representation of your surface again, right click on the surface name within the "Terrain Model Explorer", and choose the menu "Surface Display→Quick View". Now you see the problem that is driving the error in the volume computation. There are several (four) spikes which run straight up. You know from your familiarity of the site that these spikes do not exist. You wish to see the tips of these spikes, but each time you pan the drawing, the vectors disappear.

- Users have the ability to import these vectors as drawing entities and edit them, so we will take advantage of this feature. Close the "Terrain Model Explorer" if it is open and from the Land Desktop pull down menu, select "Terrain→Edit Surface→Import 3D Lines". When prompted, you may erase the old surface lines. There is nothing to erase anyway since there aren't any surface lines in this drawing yet. It is not always necessary to import vector lines and it is actually discouraged unless users need to make surface edits. The imported surface entities take up unnecessary space and are not needed to perform Land Desktop computations, but we will leave them in the drawing for this assignment.

- If the "Prototype:" "Default (Feet)" was used when creating the project as instructed earlier in this lesson, then the surface entities will have been placed in the drawing on the layer "SRF-VIEW". Zoom extents so that you are able to see the entire surface. Now the four spikes are more evident.

- We want to see which points these spikes are going to in plan view. In order to identify them set the layer "C-CNST-LINE" current and using your endpoint OSNAP connect the tips of the 4 spikes with AutoCAD lines.

- Save this view by typing "V→New" and giving it a name "ISO" and selecting "OK" twice to exit the dialogue box. Now go to the plan view by typing "Plan" at the command prompt and pressing the "Enter" key twice. Notice where your construction line is in plan view.

- Freeze the layer that your surface vectors are on for the time being and zoom in on the most North Westerly point that your construction line makes contact with. A display of this project point will indicate a bad elevation of "6562.00". Correct this elevation to read "129.26". If you can't remember how to edit point elevations refer to assignment number 6.

- Go into the terrain model explorer and rebuild your surface by right selecting the surface name and choosing "Build". Change your view by typing "V→ISO→Set Current→OK". The "View Manager" saves the layer state with the view by default. Therefore, the previous set of vector lines should be visible. If vector lines are not present, then thaw this layer. Re-import your vector lines ("Terrain→Edit Surface→Import 3D Lines") to see that the spike to the left is gone.

- Although you know that the other three spikes are wrong as well, you don't have elevations for them, so you choose to remove this data from the surface but not from the project points database. Remove the remaining three bad vertices using your endpoint OSNAP and the Land Desktop command

"Terrain→Edit Surface→Delete Point". Your object snaps will not work if Land Desktop is prompting for "Point Numbers". If Land Desktop is prompting for "Point Numbers", then use the toggle ".P" to request that the software prompt for "Point to delete". This command does not delete project Cogo points. It merely removes surface vertices. Although the software is supposed to regenerate surface vectors as vertices are removed from the surface, this doesn't always happen. You can be assured that surface vertices have been removed if subsequent attempts cause Land Desktop to issue the prompt: "Error - Unable to delete specified point.".

- Save your surface by selecting "Terrain→Save Current Surface". All of the bad surface vertices are gone and your surface is correct. Re-import the surface vectors and see for yourself.

- Export the project points to the "Survey" directory of the corresponding project with the name "Export.txt" using the same parameters that you used to import the text file initially.

- Isolate your surface vectors and viewport layer. While on the "C-ANNO-VPRT" layer, create a 40 scale landscape 24′ × 36′ floating viewport of the isometric view.

- Your global linetype scale for paperspace plotting should be set to "1". The AutoCAD system variable "PSLTSCALE" should also be set to "1".

- Save and exit this drawing. You have successfully completed this assignment.

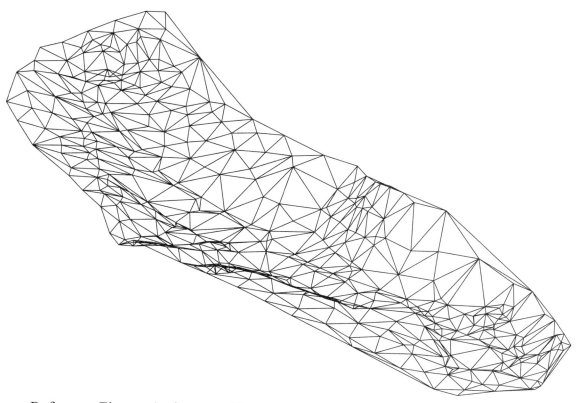

Reference Figure: Assignment 12

NOTES:

> Assignment #13
>
> ## *Faults (Breaklines)*

Recommended Assignments Prior to Working this Assignment:

Assignments 1-7, 12

Required Assignments Prior to Working this Assignment:

None

Goals and Objectives

In our last lesson, we learned how to add the third dimension to our topographic data. The third dimension that we created was basically a surface generated by the faces of several irregularly networked triangles. In this assignment, we will learn how to streamline our data through the use of faults.

Faults, also referred to as breaklines, are saved with a surface and used to create barriers in the surface which redirect triangulation. A correctly defined breakline will only allow triangulation across the breakline at its vertices. By defining a feature as a fault, Land Desktop makes each line segment of the breakline an edge of a triangle. This allows for a more accurate means of data interpretation. Areas where a user might want to construct breaklines include roadways, streams, ridges (top and toe of bank) and walls (including curbs).

The three basic types of faults are standard faults, wall faults and proximity faults. Standard faults and proximity faults are very similar so let's begin by discussing how a wall fault works.

Wall faults are generally used to simulate retaining walls and curbs. When a user chooses to create a wall fault, Land Desktop creates a three dimensional polyline with a minute offset.

Land Desktop uses linear interpolation to generate elevations between surface vertices. Linear interpolation between 2 points having the same horizontal position is mathematically impossible since zero is indivisible, so Land Desktop merely creates minute offset to simulate a vertical face. The user does not specify the offset.

By definition, the control side of the wall is the side of the wall that does not move and is opposite the offset side of the wall. The wall offset is so small, that it makes little difference which side is the physical control side of the wall, so the control side of the wall is determined by how the user wishes to define the wall. As a good rule of thumb, if a wall is defined from top to bottom, then the offset side should be down slope. In this situation, elevations are first entered for the top of wall. The user is prompted for a choice of either entering an incremental distance to be specified as a negative number or the absolute wall height at the bottom of the wall.

Standard and **Proximity** faults are generally used to simulate roadways, streams and ridges. While standard faults utilize entity and point component data for obtaining elevation information, proximity faults read their elevation information from the surface. When a proximity fault is drawn, it is merely a 2 dimensional LWPolyline. Elevation information isn't actually extracted until the surface which the fault is affiliated is built.

Exercise Instructions

- Logon to your workstation and begin a session of Land Desktop. If Land Desktop has been configured on your workstation to display the "Start Up" dialogue box, then cancel this feature so that the AutoCAD model space environment is displayed. If launching Land Desktop brings you directly into the AutoCAD model space environment, then Land Desktop has been configured to begin without the startup dialogue. If the AutoCAD Map "Task Pane" is shown, then close this dialogue box as well. Users that have the ability to customize their profiles can turn these features off permanently by following the procedure described in Assignment #1.

- Begin a new drawing by selecting "File→New". Choose to create a new project by picking the button labeled "Create Project". Select the "Prototype:" "Default (Feet)" from the drop down list. Give the project the "Name:" "Assign13-##" where the "##" reflects the number which was assigned to you on the first day of class. If the number assigned to you is a single digit (for example 4 as opposed to 14), then enter a zero preceding the single digit (such as 04). Write "Breaklines" in the "Description:" and "Keywords:" areas. This will serve as a project summary which may be used to find or filter projects using the Project Manager.

- Select "OK" to bring you back to the "New Drawing" dialogue box and type the "Name:" "Assign13-##" where the "##" reflects the number which was assigned to you on the first day of class. Make sure that the "Project Name:" displays the correct project.

- Select the "Acad.dwt" template and then "OK" to generate a new drawing in the "DWG" folder of the current project. After selecting the "OK" button, you may be prompted to save changes to the previous drawing session. Since there were no objects in the previous drawing session worth saving, choose not to save changes.

- You will be prompted with the "Load Settings" dialogue box which displays a list of drawing setups to choose from. These setups are saved back to the path determined by the Network Administrator who installed the software. The default path is to the local machine and is displayed in the dialogue box. We will set up our parameters manually, so select "Next" to set up your project with the following parameters:

Units Area
  Linear Units = Feet
  Angle Units = Degrees
  Angle Display Style = Bearings
  Display Precision Linear = 2
  Display Precision Elevation = 2
  Display Precision Coordinate = 5
  Display Precision Angular = 4
  "Next"

Scale Area
  Horizontal = 10
  Vertical = 2
  Paper Size = 24 x 36(D)
  "Next"

Zone Area
  "Next"

Orientation Area
  "Next"

Text Style Area
  Leroy.stp
  L80

- Then select "Finish" and a screen will display providing the user with a summary of the settings chosen. Review the settings and select "OK". Although users have the ability to save settings for retrieval with future projects, assignments in this text require that you set up the parameters manually for practice in each project.

- You will be prompted with the "Create Point Database" dialogue box. Accept the default settings by choosing "OK".

- Create several new layers with the following parameters:

Layer	Color	Linetype
C-ANNO-VPRT	White	Continuous
C-FLTS-FL	Cyan	Dashdot2
C-FLTS-TOP	Yellow	Hidden2
C-FLTS-WALL	Blue	Continuous
C-PNTS-EG	Green	Continuous
C-PNTS-FL	Red	Continuous
C-PNTS-TOP	Green	Continuous
C-PNTS-WALL	Red	Continuous

- Set your point settings so that your point text comes in with the L80 text style. Using the "+" marker, your marker height should be set at ½ the model space text height. The corresponding marker height for a text height of 0.80 units is 0.40. Set a text rotation angle of 30 degrees and set you points up so that they are not inserted to the drawing as they are imported into the project.

- Import your points to the project COGO points database without inserting the points into the drawing. Lesson files may be downloaded from www.schroff1.com. The file makes use of the comma as a delimiter and takes on the format "PNEZD".

- You created one "C-PNTS-XXXX" layer for each description type (i.e. EG, FL, TOP & WALL). While on the correct layer, insert points into your drawing. If you forgot how to accomplish this, refer to Assignment 5. These points represent the vertices of the new surface that you are about to create.

- Build a surface named "EG_NoFLTS" from all of the project points.

- Build a second surface from the same data named "EG_FLTS".

- Import 3D vector lines from one of the two surfaces as you did in the last lesson. Enter the layer dialogue box, change the layer name from "SRF-VIEW" to "C-SURF-VIEW-FLTS-NONE" and change its color to green. Now this set of 3D vectors will not be deleted the next time the vectors are imported.

- Isolate the layers "C-PNTS-TOP" and "C-FLTS-TOP". While on the layer "C-FLTS-TOP", proceed to connect the 2D nodes with LWPolylines (This command may be executed by typing "PL" at the command line.) as shown in the figure at the end of this chapter. These LWPolylines will later be used to construct proximity faults.

- Isolate the layers "C-PNTS-FL" and "C-FLTS-FL". While on the layer "C-FLTS-FL", proceed to connect the 2D nodes with an LWPolyline as shown in the figure at the end of this chapter.

- Isolate the layers "C-PNTS-WALL" and "C-FLTS-WALL". While on the layer "C-FLTS-WALL", proceed to connect the 2D nodes with an LWPolyline beginning with point number 14 as shown in the figure at the end of this chapter.

- Isolate the layer "C-FLTS-FL". Defining a proximity fault for the creek flow line will be our next task. Launch the terrain model explorer. Click the "+" symbol to the left of the surface named "EG_FLTS" to view the data sets. Right click on the breaklines and elect to define proximity faults by selecting

"Proximity By Polylines". Name the breakline "FL" and select the flow line when prompted. Choose not to delete the entities when prompted. If the correct prototype was used as instructed earlier in this assignment when creating this project, a 2-dimensional fault will have been created in the model space environment for the creek flow line on the layer "SRF-FLT"

- Define the proximity faults for the top of bank in the same manner that a proximity fault was created for the creek flow line. Both top of bank faults can be given the same name and defined simultaneously.

- While in the terrain model explorer viewing the data sets, right click on the "Breaklines" data set and select "List Breaklines". A list of all defined breaklines will appear. Pick one of the proximity breaklines and then choose the "List" option. You will find that Land Desktop has not yet assigned elevations to the breakline. Exit the terrain model explorer.

- Isolate and zoom in on the retaining wall. Launch the terrain model explorer and right click on the "Breakline" data set and select "Define Wall Breaklines". When asked for a wall description, name the breakline "WALL". Select the polyline when prompted. The offset side will be the down slope side of the wall. In this case the hill slopes down to the East. We can see from listing the project point data that the top of the wall at point No 14 has an elevation of 138.52ft. Enter the elevation 138.52 for the top of the retaining wall. When prompted with "Elevation (or Difference)", type "D" to specify the distance in feet to the bottom of the retaining wall. A set of field notes reveals that the survey elevation was taken at the top of a 5ft tall retaining wall. Enter a value of "-5" to instruct Land Desktop that the bottom of the wall is 5ft below the surveyed elevation you just keyed in. The software will then prompt for the elevation at the top of wall for the next surveyed elevation. The top of the wall at points 15 and 16 are also identified by their respective point elevations. The entire wall height is 5ft, so use a difference of "-5" to arrive at the bottom of the retaining wall for each subsequent surveyed elevation. Proceed to define the wall at the remaining two vertices. Notice that your polyline has been modified and changed to the layer "SRF-FLT". A couple of "Enters" will bring you back to the command line.

- Let's have a look at the wall fault in three dimensions. Type "VP" at the command line and enter an angle of 290 degrees from the x-axis and 30 degrees from the xy- plane. Then select "OK". Zoom in and examine your 3 dimensional polyline on the layer "SRF-FLT". Return to plan view after briefly examining the fault by typing the command "Plan" and pressing "Enter" twice.

- Rebuild your surface "EG_FLTS", but this time include project point data and fault information. This can be accomplished by turning on the check boxes labeled "Use breakline data" and "Convert proximity breaklines to standard" after electing to build the surface.

- Import 3D vector lines for the modified surface. Change the layer to name from "SRF-VIEW" to "C-SURF-VIEW-FLTS" with a color of red.

- Isolate the two sets of surface vectors and compare them. Look specifically in the areas where you defined faults. You will notice that the triangulation for the redefined surface containing faults only crosses the breaklines at the vertices in the breaklines.

- Create a 10 scale 24′ × 36′ floating viewport in plan view while on the corresponding layer. Your global linetype scale for paperspace plotting should be set to "1". The AutoCAD system variable "PSLTSCALE" should also be set to "1".

- Save and exit your drawing. You have successfully completed this project.

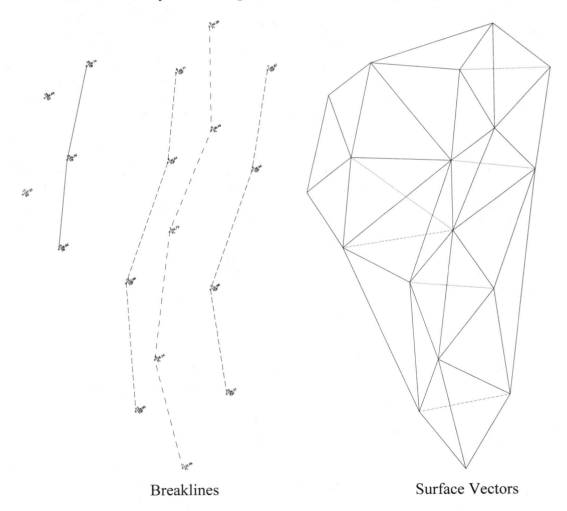

Breaklines   Surface Vectors

Reference Figure: Assignment 13

> Assignment #14
>
> ## *Creating Contours from Surface Data*

Recommended Assignments Prior to Working this Assignment:

Assignments 1-7, 12-13

Required Assignments Prior to Working this Assignment:

Assignment 13

Goals and Objectives

In our last couple of lessons, we learned how to build a surface from our project point data and construct breaklines so that Land Desktop would be able to interpret our project point data correctly. Since most civil engineers and land surveyors work with two-dimensional drawings, and the surface that we created is three-dimensional, our next task will be to display our information accordingly. One way that we can display our data is through the use of contours. Contours are a 2 dimensional representation of a 3 dimensional surface.

In this assignment, we will learn how to create contours from our surface and touch a bit more on surfaces as we implement a polyline boundary to confine the growth of our surface.

This lesson will also target contour labeling. We will see from this assignment that when entities of type "AEC Contour" are exploded, they become 2D polylines with a constant elevation (Z coordinate) equal to the value of the AEC Contour prior to its being exploded.

Exercise Instructions

- Logon to your workstation and begin a session of Land Desktop. If Land Desktop has been configured on your workstation to display the "Start Up" dialogue box, then cancel this feature so that the AutoCAD model space environment is displayed. If launching Land Desktop brings you directly into the AutoCAD model space environment, then Land Desktop has been configured to begin without the startup dialogue. If the AutoCAD Map "Task Pane" is shown, then close this dialogue box as well. Users that have the ability to customize their profiles can turn these features off permanently by following the procedure described in Assignment #1.

- Use the Land Desktop project manager (Projects→Project Manager) to copy "Assign13-##" to a new project named "Assign14-##" where the "##" reflects the number which was assigned to you on the first day of class. Refer to Assignment 4 if you can't remember how this was accomplished.

- In the "Copy Project To" "Name:" location type "Assign14-##" where the "##" reflects the number which was assigned to you on the first day of class. If the number assigned to you is a single digit (for example 4 as opposed to 14), then enter a zero preceding the single digit (such as 04). In the "Description:" and "Keywords:" areas, enter the text "Contours". Choose "OK" and Land Desktop will ask you if you want to reassociate the copied drawing files so that the copied drawings reference the copied project. Choose "Yes" and then "Close".

- We now have a drawing file named "Assign13-##" in our "Assign14-##" project. It is important that we delete this file before proceeding so that there is no confusion as to which file is the correct "Assign13-##" drawing file. Use Windows Explorer or some other file manager to delete the "Assign13-##.dwg" drawing which resides in your "Assign14-##" project.

- Begin a new drawing by selecting "File→New" to bring up the "New Drawing" dialogue box. Type the "Name:" "Assign14-##" where the "##" reflects the number which was assigned to you on the first day of class.

- We are presently working with Project: "Assign14-##". Make sure that the "Project Name:" displays the correct project, select the "Acad.dwt" template and then "OK" to generate a new drawing in the "DWG" folder of the current project. After selecting the "OK" button, you may be prompted to save changes to the previous drawing session. Since there were no objects in the previous drawing session worth saving, choose not to save changes.

- You will be prompted with the "Load Settings" dialogue box which displays a list of drawing setups to choose from. These setups are saved back to the path determined by the Network Administrator who installed the software. The default path is to the local machine and is displayed above. We will set up our parameters manually, so select "Next" to set up your project with the following parameters:

    <u>Units Area</u>
    Linear Units = Feet
    Angle Units = Degrees
    Angle Display Style = Bearings
    Display Precision Linear = 2
    Display Precision Elevation = 2
    Display Precision Coordinate = 5
    Display Precision Angular = 4
    "Next"

Scale Area
   Horizontal = 10
   Vertical = 5
   Paper Size = 24 x 36(D)
   "Next"

Zone Area
   "Next"

Orientation Area
   "Next"

Text Style Area
   Leroy.stp
   L100

- Then select "Finish" and a screen will display providing the user with a summary of the settings chosen. Review the settings and select "OK". Although users have the ability to save settings for retrieval with future projects, assignments in this text require that you set up the parameters manually for practice in each project.

- You will not be prompted with the "Create Point Database" dialogue box, because point information was copied from the "Assign13-##" project.

- Create a new layer having the following parameters:

Layer	Color	Linetype
C-ANNO-VPRT	White	Continuous

- Open the terrain model explorer and set the "EG_NoFLTS" surface current by right clicking on it and choosing the menu "Open (Set Current)". You may then close the terrain model explorer.

- Select "Terrain→Contour Style Manager→Manage Styles" to create a new contour style. Add the contour style by typing the name "CONT-L100-010X" in the area labeled "Contour Styles in Drawing" and selecting the "Add" button. Go to the "Text Style" tab of the contour style manager and specify to use the L100 text style. Change the precision to "0". Go to the "Contour Appearance" tab and select the option to "Add Vertices". Set smoothing value to 5 and exit the contour style manager by selecting "OK".

- Use the command "Terrain→Create Contours" to construct contours from the "EG_NoFLTS" surface having a minor elevation of 1ft and a major elevation of 5ft. Use a vertical scale factor of 1. Allow the contours to come in on their default layers. A graphic is shown on the following page for reference.

Introduction to Land Desktop 2007

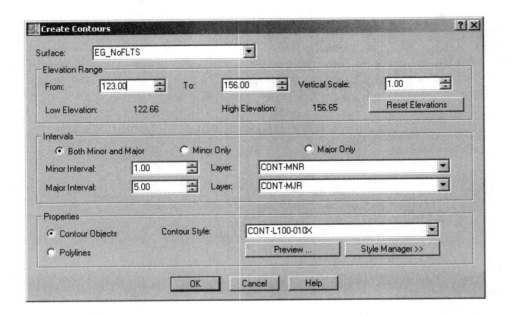

After contours are created, go in to the layer dialogue box and change the properties of the layers as follows:

Existing Layer Name	New Layer Name	New Color
CONT-MJR	C-CONT-MAJR-FLTS-NONE	Red
CONT-MNR	C-CONT-MINR-FLTS-NONE	245

- List one of the contours. Notice how AutoCAD informs the user that the entity is an AEC contour. Now elect to explode one of the contours one time and list it again. This time AutoCAD informs the user that the entity is an LWPolyline. Also take a look at the number of vertices that have been added to the contour that was exploded. Undo the explode so that the contour is once again an AEC contour. List the previously exploded contour to ensure that it has been changed back to an AEC contour.

- Insert the assignment 13 drawing file as a block at the coordinates (0,0) with a scale factor of one and "0" rotation angle. Explode the drawing one time. If you elected to have the drawing exploded using the check box in the dialogue box upon insertion, then do not explode the drawing a second time.

- Isolate the "C-PNTS-", "C-FLTS-" and "C-CONT-" layers. Given what you know about the site, the contours should not appear even remotely correct. There is no evidence of the retaining wall, and the creek is nonexistent in several locations. Refer to the graphic on the following page for reference.

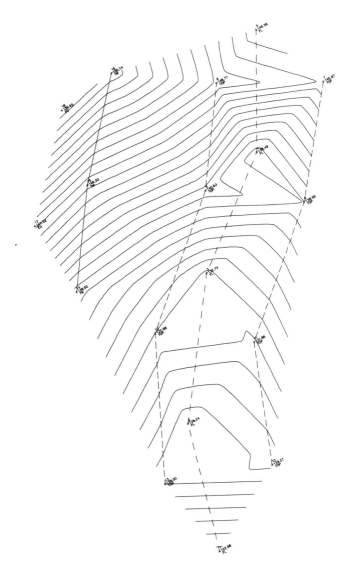

- Use the command "Terrain→Create Contours" to construct contours from the "EG_FLTS" surface with a minor elevation of 1ft and a major elevation of 5ft. Use a vertical scale factor of 1. Allow the contours to come in on their default layers. After the contours are created, go in to the layer dialogue box and change the properties of the layers as follows:

Existing Layer Name	New Layer Name	New Color
CONT-MJR	C-CONT-MAJR-FLTS	Green
CONT-MNR	C-CONT-MINR-FLTS	Blue

- Isolate the "C-PNTS-", "C-FLTS-" and "C-CONT-" layers. Notice how the contours that were generated from the surface having breaklines deviate from the previous set of contours and that these contours do reflect the site geometry. This is demonstrated in the reference figure at the end of this assignment.

- Insert the AutoCAD drawing file which corresponds to this lesson at the coordinates (0,0) with a scale factor of one and "0" rotation angle. Lesson files may be downloaded from www.schroff1.com. Explode the drawing one time. If you elected to explode the drawing upon inserting, then do not explode it a second time.

- Launch the "Terrain Model Explorer" and expand the TIN data types by picking the "+" symbol adjacent to the surface name "EG_FLTS". Right select on "Boundaries" and then select "Add Boundary Definition". Select the polyline in the AutoCAD environment which you just blocked in. Toggle through the options accepting the defaults by pushing "Enter" on the keyboard 4 times to add boundary definition. If you get an error message "Can only work with polylines", then the block that you inserted may have been exploded too many times.

- Rebuild your surface and import the new vector lines. Notice how the surface growth has been limited to the polyline boundary.

- Isolate the contours which you generated from the surface having faults.

- Generate contour labels using the command "Terrain→Contour Labels→Group Interior". When the dialogue box comes up, set the "Elevation Increment" to 5. By setting this value to 5, we are telling Land Desktop to label every contour having an elevation which may be divided by 5 to obtain a whole number. Turn on the check box to "Add multiple interior labels along each contour" every 50ft. The command labels contours by instructing users to pick two points. Every contour that crosses the line made up by the two points picked which meets the criteria specified in the dialogue box obtains a label. Pick your first point just south of the contours. Pick your second point just north of the contours.

- Isolate the "C-PNTS-", "C-FLTS-" and "C-CONT-" layers and create a 24′×36′ 10 scale landscape floating viewport in plan view on the corresponding layer. Your global linetype scale for paperspace plotting should be set to "1". The AutoCAD system variable "PSLTSCALE" should also be set to "1".

- Save and exit your drawing. You have successfully completed this assignment.

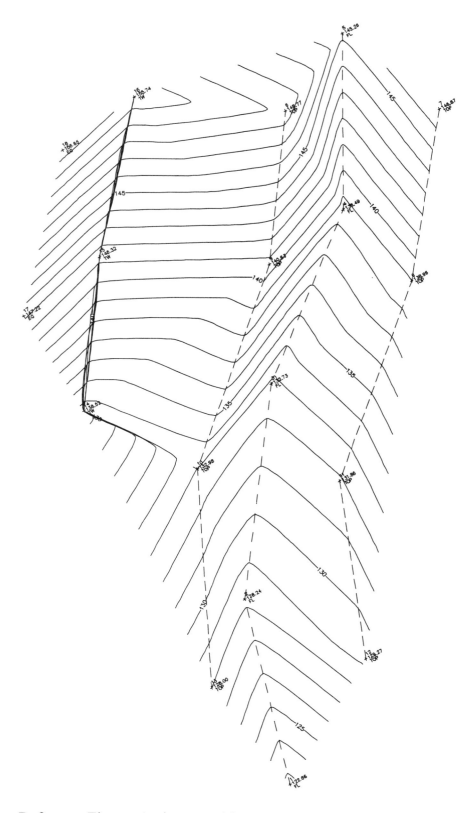

Reference Figure: Assignment 14

NOTES:

Assignment #15

Land Desktop Cross Sections

Recommended Assignments Prior to Working this Assignment:

Assignments 1-7, 12-14

Required Assignments Prior to Working this Assignment:

None

Goals and Objectives

A cross section is a drawing of an object or objects cut lengthwise to show the vertical interior makeup. Sections are used extensively by civil engineers to help determine what is happening in a given plane and to convey this information to others. Land Desktop allows users to quickly produce cross sections through defined surfaces.

In this assignment we will create two new surfaces and explore two methods for constructing cross sections through them. In one instance we will explore the Land Desktop Section View to get a streamlined quick look at the cross section and in the other we will define, process and import sections to generate editable linework.

Exercise Instructions

- Logon to your workstation and begin a session of Land Desktop. If Land Desktop has been configured on your workstation to display the "Start Up" dialogue box, then cancel this feature so that the AutoCAD model space environment is displayed. If launching Land Desktop brings you directly into the AutoCAD model space environment, then Land Desktop has been configured to begin without the startup dialogue. If the AutoCAD Map "Task Pane" is shown, then close this dialogue box as well. Users that have the ability to customize their profiles can turn these features off permanently by following the procedure described in Assignment #1.

- Begin a new drawing by selecting "File→New". Choose to create a new project by picking the button labeled "Create Project". Select the "Prototype:" "Default (Feet)" from the drop down list. Give the project the "Name:" "Assign15-##" where the "##" reflects the number which was assigned to you on the first day of class. If the number assigned to you is a single digit (for example 4 as opposed to 14), then enter a zero preceding the single digit (such as 04). Write "Land Desktop Cross Sections" in the "Description:" and "Keywords:" areas. This will serve as a project summary which may be used to find or filter projects using the Project Manager.

- Select "OK" to bring you back to the "New Drawing" dialogue box and type the "Name:" "Assign15-##" where the "##" reflects the number which was assigned to you on the first day of class. Make sure that the "Project Name:" displays the correct project.

- Select the "Acad.dwt" template and then "OK" to generate a new drawing in the "DWG" folder of the current project. After selecting the "OK" button, you may be prompted to save changes to the previous drawing session. Since there were no objects in the previous drawing session worth saving, choose not to save changes.

- You will be prompted with the "Load Settings" dialogue box which displays a list of drawing setups to choose from. These setups are saved back to the path determined by the Network Administrator who installed the software. The default path is to the local machine and is displayed in the dialogue box. We will set up our parameters manually, so select "Next" to set up your project with the following parameters:

    Units Area
        Linear Units = Feet
        Angle Units = Degrees
        Angle Display Style = Bearings
        Display Precision Linear = 2
        Display Precision Elevation = 2
        Display Precision Coordinate = 5
        Display Precision Angular = 4
        "Next"

    Scale Area
        Horizontal = 20
        Vertical = 5
        Paper Size = 24 x 36(D)
        "Next"

    Zone Area
        "Next"

    Orientation Area
        "Next"

    Text Style Area
        Leroy.stp
        L100

- Then select "Finish" and a screen will display providing the user with a summary of the settings chosen. Review the settings and select "OK".

Although users have the ability to save settings for retrieval with future projects, assignments in this text require that you set up the parameters manually for practice in each project.

- You will be prompted with the "Create Point Database" dialogue box. Accept the default settings by choosing "OK".

- Create a new layer with the following parameters:

Layer	Color	Linetype
C-ANNO-VPRT	White	Continuous

- Insert the drawing file corresponding to this project at the coordinates 0,0 with a scale of 1 and with a rotation angle of 0. Lesson files may be downloaded from www.schroff1.com. Zoom extents and explode the drawing one time.

- View the objects in 3D to get a feel for what is going on (this can be accomplished by typing "VP" at the command line and entering an angle of 260 degrees from the X Axis and 25 degrees from the XY Plane) and then go back to the plan view.

- Isolate the layers "C-CONT-MINR-TOPO" and "CONT-MAJR-TOPO" and build a surface from these contours. You can accomplish this by following the steps shown below.

    1. "Terrain→Terrain Model Explorer".
    2. "Manager→Create Surface". A new surface titled "Surface1" will have been created in the "Terrain" folder. Right click on this new surface and choose to rename it to "EG".
    3. Right click on the newly named "EG" surface and select "Properties". On the "Surface" tab of the "Properties" dialogue box, enter a description which reads "Existing Ground from topographic contours". Place a check mark in the box adjacent to the text which reads "Minimize flat triangles resulting from contour data". Select "OK" to exit the "Properties" dialogue box.
    4. Make sure that the "Terrain" menu to the left is expanded. It is expanded if a "-" symbol appears to its left. If a "+" symbol appears to the left of the "Terrain" text, click the "+" symbol to expand the surfaces. Click the "+" symbol to the left of the surface name "EG" to view the surface data types. Right click on contours to "Add Contour Data". Accept the defaults for "Contour Weeding" and use the "Entity" option to select the isolated contours which appear on your screen.
    5. Right click on the surface "EG" and "Build" the surface. Accept the defaults in the "Build Surface Dialogue" box and "OK" to the message indicating that the software is done building the surface. You may exit or minimize the Terrain Model Explorer.

- Now isolate the layer "C-CONT-MOUND" and create a new surface called "Mound" from this contour data using the same procedure that was used to create a surface from the topographic map contours.

- Turn on all of the layers and zoom in on the section lines in your drawing labeled A-A, B-B and C-C. We have just created surfaces from our contours, and are about to create cross sections from the surface data. At this point, our contours are no longer needed and will get in the way, so you may want to turn them off. This can be accomplished by isolating the layers "C-ANNO-SCTN-LINE" and "C-ANNO-TEXT-SCTN".

- Select "Terrain→Sections→View Quick Section" and pick the base line for Section A-A. Press "Enter" one time and a cross section view will pop up on your screen. This method for generating cross sections is generally for the user to get a feel for what is happening in the area of concern. You will notice that the viewer only shows a cross section through the active surface. In this case the active surface is the "Mound" surface which was the last surface created. Move your quick view section aside so that you are able to see your cross section line A-A. Select the line and then "Right Click→Flip Direction". Notice that your section flips direction. This is an indication that the cross section direction is dictated by the direction of the section line in plan view. Close your section view by clicking on the "X" in the upper right hand corner of the section view.

- Toggle "On" the multiple surfaces by selecting "Terrain→Sections→Multiple Surfaces On/Off". You will see that this menu works as a toggle. If multiple surfaces are off then running the routine turns them on. If multiple surfaces are on, then running the routine turns them off again. Run the routine a couple of times so that you get a feel for how it works. Leave them toggled on when are finished experimenting.

- Select your multiple surfaces through the menu "Terrain→Sections→Define Multiple Surfaces" and selecting each of the surfaces from the list on the left side of the dialogue box while holding down the "Ctrl" key. Holding down the control key allows users to pick more than one surface. Press "OK" to exit the dialogue box.

- Create another section view along section A-A using "Terrain→Sections→View Quick Section" and pick the base line for Section A-A. Notice that two surfaces appear this time. Scales, grid, color and other view properties can be changed for the section views by selecting "Section→View Properties" from within the section view editor. Slope and elevation statistics may be obtained through the "Utilities" menu within the section view editor.

- We will next import this cross section into the AutoCAD environment as objects so that we have the opportunity to work with it. From within the section view editor, select "Utilities→Import Quick Section". Choose not to enter a "Layer name prefix" when prompted by accepting the default. Also accept the default for the "Datum" layer. When prompted for a description enter "A-A". When prompted for an insertion point, enter the coordinate pair (15000,5400). Assign a datum elevation of 110.

- Exit the "Quick Section Viewer" and zoom extents. Enter the layer dialogue box to change the color of the "Mound" layer to green and the color of the "EG" layer to Red.

- The "Quick Section Viewer" is a handy tool to implement if you don't know what your surface looks like and want a quick peek before importing the section into the drawing environment. This feature can also be launched by selecting an AutoCAD line or Polyline and then right selecting and choosing "View Quick Section". If it is not desirable to view a section in the "Quick Section Viewer" before importing into the drawing environment, then users have that option of skipping this step. Simply select the line or polyline for which you desire to import a section, right select and choose "Import Quick Section" from the screen menu. This feature is also available in the "Terrain→Sections" menu. Users can generate a 3-dimensional polyline which follows the surface along a given section line by selecting the cross section line, right clicking and choosing the menu "Project Object". Although these last few tips are not required for this assignment, feel free to experiment with them.

- The next method for determining cross sections is commonly used to generate cross sections for a hard copy plot. Exit existing section views which you may have open by selecting the "X" box in the upper right hand corner of each of the boxes.

- Select "Terrain→Sections→Define Sections". Give the sections the group label name "Mound". Name the first section label "B-B". After an "Enter", use your object snaps to pick the end points of your section line for the B-B section from left to right. You may need to use the transparent (') zoom command to accomplish this. Proceed to define section "C-C" in the same manner. "Enter" twice to terminate the command.

- Select "Terrain→Sections→Process Sections" to process the data.

- Select "Terrain→Sections→Import Sections" to bring your processed sections into the drawing environment. Accept the default layer datum and the vertical scale factor of 4 but make the datum elevation for each of your cross sections elevation 110. [The default scale factor comes from the quotient of the horizontal and vertical scales (20/5)]. The user will be prompted to select a screen location for the cross sections. Enter the coordinate pair (15000,5600) for section B-B and (15000,6000) for section C-C. It is generally a good idea to use a common datum when possible for all of your cross sections to minimize mistakes made while reading data from them. It is also a good idea to line up the sections to make it easier for users that have to work with them.

- Select "Terrain→Sections→Grid for Sections" to place a grid on each of the sections. Accept the default layer "Grid" and select the datum block for Section A-A. The datum block is the white block to the left of the section entities. Type an elevation increment of 5 and an offset increment of 20 when prompted. Proceed to place grids on the remaining two cross sections using the same parameters. Change the color of the grid layer to 245.

- Create a series of 20 scale floating viewports in plan view on the corresponding layer. The viewports should be sized to fit inside a 24′ × 36′ sheet of paper. This means that the lower viewports should be 15′ tall and the upper viewport should be 9′ tall. Use a "Dview Twist" to alter the orientation of the objects within the plan view viewport. Do not attempt to move your cross sections in the model space environment.

- Your global linetype scale for paperspace plotting should be set to "1". The AutoCAD system variable "PSLTSCALE" should also be set to "1".

- Save and Exit your drawing. You have successfully completed this project.

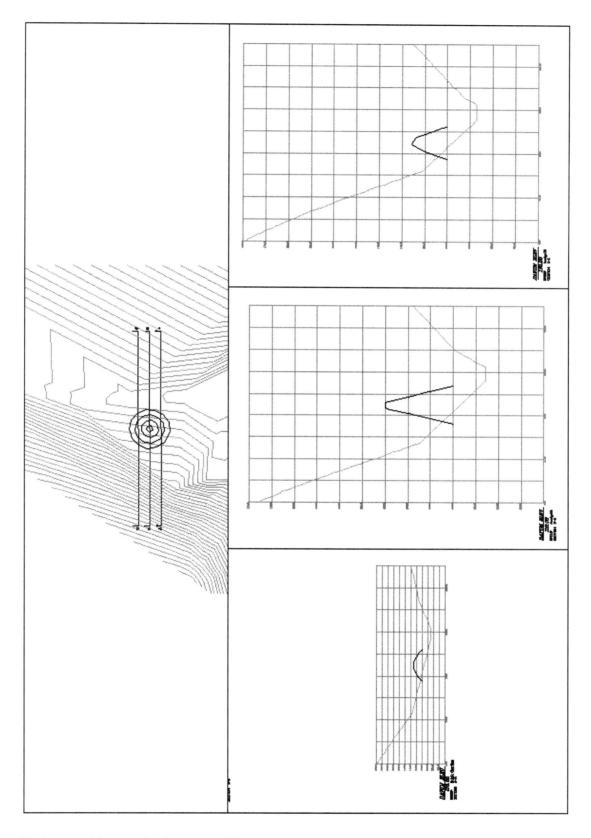

Reference Figure: Assignment 15

NOTES:

Assignment #16

Earthwork Volumes

Recommended Assignments Prior to Working this Assignment:

Assignments 1-7, 12-15

Required Assignments Prior to Working this Assignment:

None

Goals and Objectives

You may have learned somewhere along the line in your advanced AutoCAD training that volumes are obtainable between planes if the planes make up the faces of a solid. For instance, the volume of a solid box can be obtained through AutoCAD's "Tools→Inquiry→Mass Properties" command. In this assignment, we will learn how Land Desktop gives users the ability to create volumes from our surfaces and compare the result to the volume generated by the AutoCAD mass properties routine.

Land Desktop gives users three general methods for computing volumes from defined surfaces. The three methods are the Grid, Composite Surface, and the Section (which can be performed both with Average End Area and Prismoidal). Before we learn about the differences between the three methods, it might be useful to discuss some of the Land Desktop nomenclature.

"Stratum" identifies two surfaces that exist in your project for which users wish to compute the volume. The surfaces are referred to as "Surface 1" and "Surface 2" and are commonly defined as "EG" (existing ground) and "FG" (finished ground).

A **"Site"** contains information bounding the area that the user wishes to perform the computation in. During the site definition, the user is prompted for grid "M" and "N" values. These values merely represent the "X" and "Y" spacing for the grid cells. The grid size should reflect the user's data. Grids that are much larger or smaller than the distance between surface data may yield volumes that are not representative of the true surface. As an example, if your surface has been constructed from spot elevations with separations averaging 25ft, then your grid spacing should be less than, but close to 25ft. Further grid spacing will provide erroneous results, and grid spacing that is too dense (0.5ft) may take too long to process.

The grid and section methods both rely on the "M" and "N" parameters in the site definition. The **"Grid"** method lays a mesh over each of your surfaces defined in the stratum and computes the difference in elevation between the meshes at the grid cell intersections. The **"Section"** method calculates cross sections from the two surfaces of the current stratum and generates volumes using average end area or prismoidal. Sections can be interpolated in either the M or N direction with spacing based on the grid spacing of the defined site.

Unlike the section and grid methods, the **"Composite Surface"** method does not rely on the "M" and "N" parameters in the site definition. It does however use the site boundary to confine the growth of the composite surface. The composite surface method is often the most accurate since it obtains its information directly from the surfaces defined in the current stratum. The routine creates a third surface equal to the mathematical difference between the vertices of the two TIN's (Triangular Irregular Network) defined in the stratum.

It is generally good practice to run volume computations using more than one method. It serves as a double check of your result, and it encourages the user to think about the computation.

Land Desktop users should be aware of one of the shortcomings of the "Section Method" that was not present in Softdesk 8. When using the section method, the average end area and prismoidal methods will yield varying results depending on the site definition when the site is larger than the surfaces. When the site is larger than the surfaces, surface data may not begin at the section corresponding to station 0. This may cause a problem, because the average end area method begins at station 0 when calculating volumes. The volume results vary depending on how far the 0 section line (datum of site definition) is from the surface data. The further away, the more impacted volumes will be.

You will see from this assignment and future experience that keeping track of your strata, sites and composite surfaces can be quite cumbersome. It's a good idea for beginning operators use common names when working with these elements. You will find that giving a single stratum, site name and composite surface a common name as "CompX" will simplify your volume computation experiences.

Exercise Instructions

- Logon to your workstation and begin a session of Land Desktop. If Land Desktop has been configured on your workstation to display the "Start Up" dialogue box, then cancel this feature so that the AutoCAD model space environment is displayed. If launching Land Desktop brings you directly into the AutoCAD model space environment, then Land Desktop has been configured to begin without the startup dialogue. If the AutoCAD Map "Task Pane" is shown, then close this dialogue box as well. Users that have the ability to customize their profiles can turn these features off permanently by following the procedure described in Assignment #1.

- Begin a new drawing by selecting "File→New". Choose to create a new project by picking the button labeled "Create Project". Select the "Prototype:" "Default (Feet)" from the drop down list. Give the project the "Name:" "Assign16-##" where the "##" reflects the number which was assigned to you on the first day of class. If the number assigned to you is a single digit (for example 4 as opposed to 14), then enter a zero preceding the single digit (such as 04). Write "Volumes" in the "Description:" and "Keywords:" areas. This will serve as a project summary which may be used to find or filter projects using the Project Manager.

- Select "OK" to bring you back to the "New Drawing" dialogue box and type the "Name:" "Assign16-##" where the "##" reflects the number which was assigned to you on the first day of class. Make sure that the "Project Name:" displays the correct project.

- Select the "Acad.dwt" template and then "OK" to generate a new drawing in the "DWG" folder of the current project. After selecting the "OK" button, you may be prompted to save changes to the previous drawing session. Since there were no objects in the previous drawing session worth saving, choose not to save changes.

- You will be prompted with the "Load Settings" dialogue box which displays a list of drawing setups to choose from. These setups are saved back to the path determined by the Network Administrator who installed the software. The default path is to the local machine and is displayed in the dialogue box. We will set up our parameters manually, so select "Next" to set up your project with the following parameters:

    Units Area
        Linear Units = Feet
        Angle Units = Degrees
        Angle Display Style = Bearings
        Display Precision Linear = 2
        Display Precision Elevation = 2
        Display Precision Coordinate = 5
        Display Precision Angular = 4
        "Next"

    Scale Area
        Horizontal = 20
        Vertical = 5
        Paper Size = 8 x 11(A)
        "Next"

    Zone Area
        "Next"

    Orientation Area
        "Next"

    Text Style Area
        Leroy.stp
        L100

Then select "Finish" and a screen will display providing the user with a summary of the settings chosen. Review the settings and select "OK". Although users have the ability to save settings for retrieval with future projects, assignments in this text require that you set up the parameters manually for practice in each project.

- You will be prompted with the "Create Point Database" dialogue box. Accept the default settings by choosing "OK".

- Create several new layers with the following parameters:

Layer	Color	Linetype
C-ANNO-TEXT	Cyan	Continuous
C-ANNO-VPRT	White	Continuous
C-BOXX-SOLD	Red	Continuous
C-CONT-EGND	Green	Continuous
C-CONT-FGND	Yellow	Continuous

- While in model space on the layer "C-CONT-EGND" draw a polyline from the origin (0,0) to (100,0) to (100,100) to (0,100) and back to the origin again. Copy the entities to the layer "C-CONT-FGND".

- Use the command "Terrain→Contour Utilities→Edit Elevation" to change the elevation of the finished ground contour to 100 and the existing ground contour to 80. You might need to isolate each of the layers to do this correctly.

- Build a surface from each of the polylines. Name the finished ground polyline at elevation 100 "FG" and name the existing ground polyline at elevation 80 "EG".

- Type "VP" to change your view to 225 degrees from the x-axis and 5 degrees from the xy-plane. Look at your two polylines to assure that they are correct and save this view as "Iso". Then go back to your plan view.

- Select the command "Terrain→Select Current Stratum". When prompted to create a new stratum type the following:

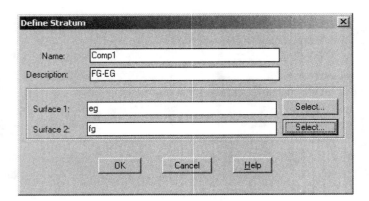

- Now that the stratum has been selected we need to define the site. Select "Terrain→Site Definition→Define Site". Enter the following:

Rotation Angle:	0
Base Point:	(-10,-10)
Msize:	10
Nsize:	10
Upper Right:	(110,110)

    Elect not to alter the size or rotation.
    Elect to erase the old site outline since there will not be any.
    Name the site "Comp1" to match the stratum.

- While in plan view, observe the site definition box that you just created and notice that it completely encompasses your contours that where used to generate surfaces for the volume computation.

- We now want to build our composite surface and allow the software to compute our cut and fill volumes. Select "Terrain→Composite Volumes"→Calculate Total Site Volume. Select the site "Comp1" from the librarian and then "OK". When prompted with the "Composite Volume Settings", accept the defaults. The **Elevation Tolerance** serves as a weeding factor in the vertical direction. Elevation differences between the surfaces identified in the stratum that are less than the identified elevation tolerance are ignored in the calculation of volumes. Cut and fill factors are used to compensate for material compaction and swell.

- When prompted for a new surface name provide "Comp1" as you did for the stratum and site name.

- A composite surface has now been computed and the volumes should have been shown on the screen.

- While on the layer "C-ANNO-TEXT" use the <u>Multiline Text Editor</u> to type the computed volume using the L100 text style on the screen in the middle of the closed polylines used to generate the surfaces.

- While on the layer "C-BOX-SOLD" use the AutoCAD "Box" command to create a box from the corner (0,0,80) to (100,100,80) with a height of 20. Use the mass properties command ("Massprop" at the command line) to compute the volume of the box. Edit the text on your screen using the "ED" command to also include the volume of the solid box. You should have come up with the same volume in different units.

- Let us now observe the impact the site definition has on volume computations when using the composite surface method. Select the command "Terrain→Select Current Stratum" and create a new stratum called "Comp2".

16.5

- Enter the following information:

Name:	Comp2
Description:	FG-EG
Surface 1:	EG
Surface 2:	FG

- Select "Terrain→Site Definition→Define Site" and enter the following:

Rotation Angle:	0
Base Point:	(25,25)
Msize:	10
Nsize:	10
Upper Right:	(75,75)

  Elect not to alter the size or rotation.
  Elect to erase the old site outline.
  Name the site "Comp2" to match the stratum.

  Notice that the site definition does not even come close to encompassing the two surfaces for which we are about to calculate volumes.

- Now build the composite surface and allow the software to compute the cut and fill volumes by selecting "Terrain→Composite Volumes"→Calculate Total Site Volume. Select the site "Comp2" from the librarian when prompted and then "OK". Accept the defaults when prompted with the "Composite Volume Settings" dialogue box. Provide the name "Comp2" for the new surface when prompted. The volume that appears at the command line should be a fraction of the previous reported volume. Edit the multiline text block again to report the most recently calculated volume.

- Let's try running a volume computation with the same stratum and site, but specifying a different "Elevation Tolerance". Select "Terrain→Composite Volumes→Calculate Total Site Volume" Select "Comp2" from the "Site Librarian" This time specify a minimum "Elevation Tolerance" of "25" In doing so, we are requesting that Land Desktop ignore elevation differences of 25ft or less when calculating volumes. Choose not to utilize a different surface. The volume that you should have obtained is zero, since the minimum elevation tolerance exceeds the vertical distance between the surfaces.

  It should be noted that in Land Desktop R2 and Softdesk products, the tolerance does not impact the volume generated. In Land Desktop R2i, R3, 2004, 2005, 2006 and 2007, the tolerance is used as a weeding factor in that any vertical distance less than the tolerance is excluded from the volume computation.

- Next we will generate a volumes report. Use the command "Terrain→ Volume Reports→Site Report→Select ALL→OK" with cut and fill factors of 1 to create a report. Select the option "Print to file" and write the data file to the "ER" folder of the corresponding project. Name the file "CALCS.TXT".

- Create a 20 scale floating viewport in plan view on the corresponding layer. Your global linetype scale for paper space plotting should be set to "1". The AutoCAD system variable "PSLTSCALE" should also be set to "1". You have successfully completed this project.

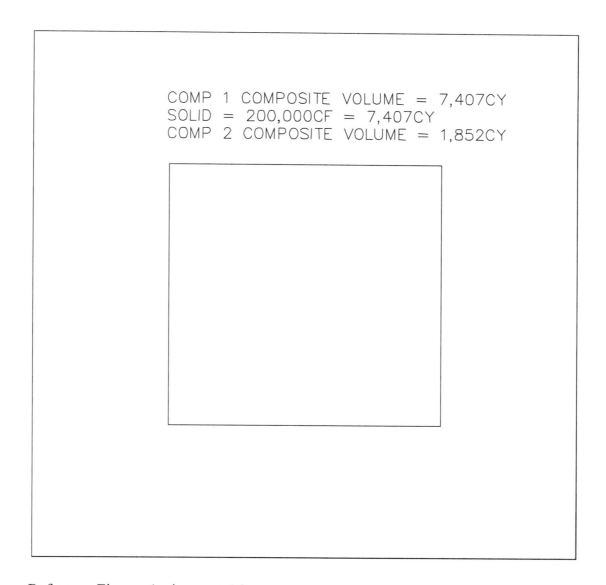

Reference Figure: Assignment 16

NOTES:

Assignment #17
Alignments

Recommended Assignments Prior to Working this Assignment:

Assignments 1-5

Required Assignments Prior to Working this Assignment:

None

Goals and Objectives

While the plan view shows what is happening in a two dimensional horizontal plane, profiles and cross sections convey information with regard to what is happening in the vertical direction. **"Stationing"** (i.e., a series of divisions indicating a distance from a point of beginning or reference) is often created so that a person reviewing a set of drawings can reference between plans, profiles and sections.

In the coming assignments we will learn how to generate profiles along a given path. The vertical information will be read from our existing ground surface. However, before creating the profile, Land Desktop will require the location of the path from which data is to be analyzed. The path that Land Desktop samples its data from is referred to as an **"Alignment"**. Defining an alignment creates an external database of information in the "Align" subdirectory for the corresponding project.

After this assignment, you will be able to define alignments by polyline and by objects (Lines and arcs). This lesson provides an introduction to Land Desktop's reporting capabilities to table stations and related information. This lesson also introduces users to alignment stationing as well as setting points at various offsets and intervals along alignments.

Your services have just been retained to stake the new road leading up to the campus computer lab. You have been provided an electronic AutoCAD file and asked by the sitework contractor to set field stakes for rough grading which identify the left and right edges of road. You have been asked to set stakes parallel to and 10ft outside of the road having a 50ft interval with reference to the roadway centerline.

Exercise Instructions

- Logon to your workstation and begin a session of Land Desktop. If Land Desktop has been configured on your workstation to display the "Start Up" dialogue box, then cancel this feature so that the AutoCAD model space environment is displayed. If launching Land Desktop brings you directly into the AutoCAD model space environment, then Land Desktop has been configured to begin without the startup dialogue. If the AutoCAD Map "Task Pane" is shown, then close this dialogue box as well. Users that have the ability to customize their profiles can turn these features off permanently by following the procedure described in Assignment #1.

- Begin a new drawing by selecting "File→New". Choose to create a new project by picking the button labeled "Create Project". Select the "Prototype:" "Default (Feet)" from the drop down list. Give the project the "Name:" "Assign17-##" where the "##" reflects the number which was assigned to you on the first day of class. If the number assigned to you is a single digit (for example 4 as opposed to 14), then enter a zero preceding the single digit (such as 04). Write "Alignments" in the "Description:" and "Keywords:" areas. This will serve as a project summary which may be used to find or filter projects using the Project Manager.

- Select "OK" to bring you back to the "New Drawing" dialogue box and type the "Name:" "Assign17-##" where the "##" reflects the number which was assigned to you on the first day of class. Make sure that the "Project Name:" displays the correct project.

- Select the "Acad.dwt" template and then "OK" to generate a new drawing in the "DWG" folder of the current project. After selecting the "OK" button, you may be prompted to save changes to the previous drawing session. Since there were no objects in the previous drawing session worth saving, choose not to save changes.

- You will be prompted with the "Load Settings" dialogue box which displays a list of drawing setups to choose from. These setups are saved back to the path determined by the Network Administrator who installed the software. The default path is to the local machine and is displayed in the dialogue box. We will set up our parameters manually, so select "Next" to set up your project with the following parameters:

    Units Area
        Linear Units = Feet
        Angle Units = Degrees
        Angle Display Style = Bearings
        Display Precision Linear = 2
        Display Precision Elevation = 2
        Display Precision Coordinate = 5
        Display Precision Angular = 4
        "Next"

    Scale Area
        Horizontal = 40
        Vertical = 10
        Paper Size = 24 x 36(D)
        "Next"

    Zone Area
        "Next"

Orientation Area
"Next"

Text Style Area
Leroy.stp
L100

- Then select "Finish" and a screen will display providing the user with a summary of the settings chosen. Review the settings and select "OK". Although users have the ability to save settings for retrieval with future projects, assignments in this text require that you set up the parameters manually for practice in each project.

- You will be prompted with the "Create Point Database" dialogue box. Accept the default settings by choosing "OK".

- Insert the drawing which corresponds to this lesson at (0,0) with a rotation angle of 0 and a scale factor of 1. Lesson files may be downloaded from www.schroff1.com. Zoom extents and proceed to explode (X) the drawing one time.

- Isolate the left storm drain layer "C-SDRN-CLLT". Use the command "Alignments→Define from Objects" to define the left storm drain. You will have to first select the object furthest to the West. Use your "Nearest" object snap to pick the left side of the first line object. Notice that Land Desktop places an "X" at the start of the alignment. When prompted, proceed to select all the entities which make up the alignment. When asked to "Select reference point", press the "Enter" key on the keyboard to skip this prompt and define the alignment with the following parameters:

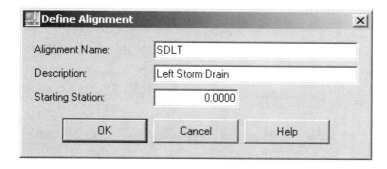

- Select "OK" when finished.

- Use the command "Alignments→ASCII file output→Output Settings" to set up the output dialogue box as follows.

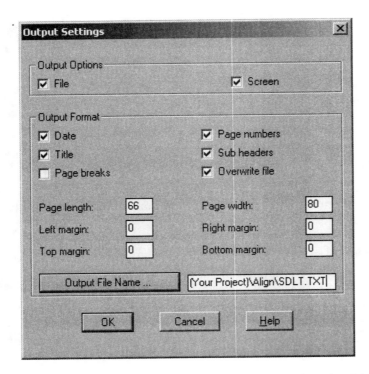

- Make sure that the output file name is saved to the "Align" directory of the corresponding project. Then select "Alignments→Edit→Station" to generate a report. Accept the start and end stations when prompted. Select "OK" when prompted until you get back to a command prompt. A text style has just been generated where specified in the output settings.

- Proceed to define an alignment and create a report for the right storm drain in the same manner which you did the left. For this Storm Drain use the following parameters:

    Alignment Name: SDRT
    Alignment Description: Right Storm Drain
    Starting Station: 0

- Give the report the name "SDRT.TXT".

- Now isolate the roadway centerline layer (C-ROAD-CTLN) and the building layer (C-BLDG-EXST). Use the command "Alignments→Define from Polyline" to define this alignment. This command will not prompt the user to window objects. The user merely has to select the single polyline.

Note: When defining your roadway centerline, you decide that you would like to maintain the roadway stationing from a current set of improvement drawings. You see from the improvement drawings that the station corresponding to the Southwest corner of the building is 5+00. In order to define your alignment appropriately, use your object snap to select the southwest building corner when prompted to "Select reference point". When prompted for a "Reference station" type "500". A dialogue box will appear as follows:

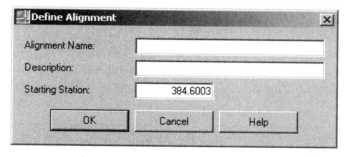

Notice that Land Desktop calculated the starting station automatically. Name the alignment "CTLN", provide a reasonable description and generate a report in the "Align" directory titled "CTLN.TXT" as was done for the alignments previously defined.

- Proceed to define alignments and create reports for the left and right hinge alignments. Since, when referencing between plan, profile and cross sections, references are made with respect to centerline stationing, it is not necessary to use a reference station when defining the hinge point alignments.

- Set your centerline alignment current using the command "Alignments→Set Current Alignment→Enter" and choosing the roadway centerline from the menu that displays.

- Set up the stationing parameters using the command "Alignments→Station Label Settings" to reflect the following:

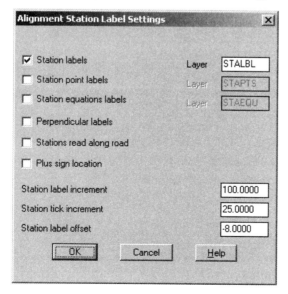

17.5

- Now label the alignment using the command "Alignments→Create Station labels". When prompted for a "Beginning station" type "400". Accept the default for the ending station which will be the end of the centerline alignment. You may choose to delete layers when prompted because stationing does not exist in this drawing.

- Use the command "Inquiry→Station/Offset Alignment" to identify the station which corresponds to the southwest building corner. Watch your command line for the results. Notice that the station is 5+00, which is the value we keyed in when using the reference option while defining the centerline alignment.

- Set your point settings as you did for Assignment 3.

- Our goal will be to set points 10ft from the edge of the road using a 50ft interval along the centerline alignment while beginning with station 4+00. For clarity make sure that the points on the left side of the road have the description "Stake-lt" and that points on the right side of the road have the description "Stake-rt". Refer to assignment 3 if you can't remember how to use automatic descriptions. While on the layer "C-PNTS-STAK", use the command "Points→Create Points-Alignments→Measure Alignment" to set your points along the left side of the road. Enter a start station of 400, an end station of 1350, an offset of -20 and an interval of 50.

- Now set stakes along the right side of road similar to the left side of the road. Keep in mind that offsets to the left of the alignment need to be keyed in as negative values while offsets to the right need to be keyed in as positive values.

- Isolate your building, centerline alignment, stakeout points, and edges of road. Create a 40 scale 24' x 36' landscape floating viewport in plan view on the corresponding layer.

- Set the global linetype scale for paperspace plotting to "1" and the AutoCAD system variable "PSLTSCALE" to "1". You have successfully completed this assignment.

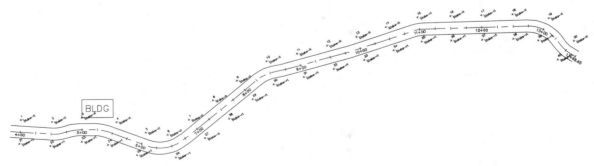

Reference Figure: Assignment 17

Assignment #18
Civil Design Profiles

Recommended Assignments Prior to Working this Assignment:

Assignments 1-7, 12-14, 17

Required Assignments Prior to Working this Assignment:

None

Goals and Objectives

A "**profile**" is similar to a cross section, because as with sections, profiles show vertical information which identifies a surfaces interior makeup. For this reason, you will find that people tend to use the terms interchangeably. Since this can lead to confusion, we will rely on the Land Desktop definitions for profiles and cross sections. A Land Desktop cross section is a view of vertical information in the transverse (perpendicular to alignments) direction. In contrast, Land Desktop profiles are views which show vertical information taken parallel to alignments.

In this assignment we will learn to create an existing ground profile using the Civil Design module in conjunction with Land Desktop, and use the vertical alignment editor to create a finished ground profile with vertical curves.

The steps to defining a profile are fairly intuitive.
- Check your profile settings to ensure that they are set to the desired values.
- Select the desired horizontal alignment.
- Select the desired surface.
- Sample data from the surface which you desire the profile to reflect.
- Create an existing ground profile.
- Create a finished ground profile.

Exercise Instructions

- Logon to your workstation and begin a session of Land Desktop. If Land Desktop has been configured on your workstation to display the "Start Up" dialogue box, then cancel this feature so that the AutoCAD model space environment is displayed. If launching Land Desktop brings you directly into the AutoCAD model space environment, then Land Desktop has been configured to begin without the startup dialogue. If the AutoCAD Map "Task Pane" is shown, then close this dialogue box as well. Users that have the ability to customize their profiles can turn these features off permanently by following the procedure described in Assignment #1.

- Begin a new drawing by selecting "File→New". Choose to create a new project by picking the button labeled "Create Project". Select the "Prototype:" "Default (Feet)" from the drop down list. Give the project the "Name:" "Assign18-##" where the "##" reflects the number which was assigned to you on the first day of class. If the number assigned to you is a single digit (for example 4 as opposed to 14), then enter a zero preceding the single digit (such as 04). Write the word "Profiles" in the "Description:" and "Keywords:" areas. This will serve as a project summary which may be used to find or filter projects using the Project Manager.

- Select "OK" to bring you back to the "New Drawing" dialogue box and type the "Name:" "Assign18-##" where the "##" reflects the number which was assigned to you on the first day of class. Make sure that the "Project Name:" displays the correct project.

- Select the "Acad.dwt" template and then "OK" to generate a new drawing in the "DWG" folder of the current project. After selecting the "OK" button, you may be prompted to save changes to the previous drawing session. Since there were no objects in the previous drawing session worth saving, choose not to save changes.

- You will be prompted with the "Load Settings" dialogue box which displays a list of drawing setups to choose from. These setups are saved back to the path determined by the Network Administrator who installed the software. The default path is to the local machine and is displayed in the dialogue box. We will set up our parameters manually, so select "Next" to set up your project with the following parameters:

    <u>Units Area</u>
       Linear Units = Feet
       Angle Units = Degrees
       Angle Display Style = Bearings
       Display Precision Linear = 2
       Display Precision Elevation = 2
       Display Precision Coordinate = 5
       Display Precision Angular = 4
       "Next"

    <u>Scale Area</u>
       Horizontal = 40
       Vertical = 10
       Paper Size = 24 x 36(D)
       "Next"

    <u>Zone Area</u>
       "Next"

<u>Orientation Area</u>
"Next"

<u>Text Style Area</u>
Leroy.stp
L100

- Then select "Finish" and a screen will display providing the user with a summary of the settings chosen. Review the settings and select "OK". Although users have the ability to save settings for retrieval with future projects, assignments in this text require that you set up the parameters manually for practice in each project.

- You will be prompted with the "Create Point Database" dialogue box. Accept the default settings by choosing "OK".

- Insert the drawing which corresponds to this lesson at the coordinate pair (0,0) with a rotation angle of 0 and a scale factor of 1. Lesson files may be downloaded from www.schroff1.com. Zoom extents and proceed to explode (X) the drawing one time.

- Create a new layer with the following parameters:

<u>Layer</u>	<u>Color</u>	<u>Linetype</u>
C-ANNO-VPRT	White	Continuous

- Isolate the two contour layers and build a surface named "EG" from the contours using the "Terrain Model Explorer". Refer to assignment 12 if you can't remember how this was accomplished. After building the surface, turn on and thaw all layers.

- Define the road centerline as an alignment from SOUTH to the NORTH beginning with the station 1+00. Refer to assignment 17 if you can't remember how this was accomplished.

- Station label your alignment by only selecting the option for "Station Labels" and set up your stationing with the following parameters:

    Label Increment: 50
    Tick Increment: 25
    Label Offset: 2

- Change "Workspaces" by selecting the Workspace "Civil Design" from the drop down list in the "Workspaces" toolbar. If the "Workspaces" toolbar is not displayed on the screen, then it can be invoked using the command "Projects→Workspaces". ("Workspaces" is a new feature that was introduced in the AutoCAD 2006 suite of products and has taken the place of the "Menu Palette Manager" used in previous versions of Land Desktop.)] Select "Profiles→Profile Settings→Values" to set the values for your profile. Set the profile values as follows:

Item	What the item does
Tangent Labels	Labels elevations along FG lines.
Vertical Curve Labels	Labels elevations along FG vertical curves.

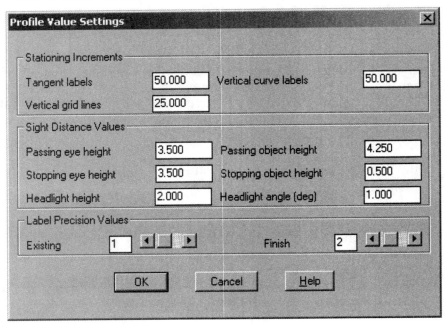

Vertical Grid Lines	Controls existing ground grid line label spacing.
Site Distance Values	Design aide for selecting vertical curve lengths

Note: Grid station labels are placed every other "Vertical Grid Line".

- Select "Alignments→Set Current Alignment", press enter once and choose the centerline alignment that you just defined a few moments ago.

- Now select "Profiles→Surfaces→Set Current Surface" and select your "EG" surface. This will open and set the "EG" surface current. (If the surface was created in this session of Land Desktop, then this step is not actually needed, because the last surface created using the terrain model explorer will be open and current.)

- Select "Profiles→Existing Ground→Sample from Surface" to sample vertical data from your current surface along your current horizontal alignment. Use an offset tolerance of .1 and make sure that the check boxes for importing sample lines and offset sampling are toggled off. You have successfully sampled data for your vertical alignment.

- Let us now create the profile. Select "Profiles→Create Profile→Full Profile" and set up your profile using the following parameters:

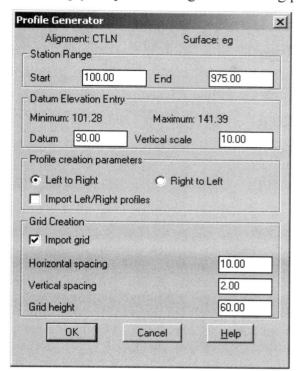

Select "OK" and insert your profile at the coordinate pair (16000,5000).

- Change the color of the layer "PEGC" by layer to green and the color of the layer "PGRID" by layer to 245. This will make the profile easier to read. Change the layer name "PGRID" to "C-PGRID-MINR". Study the profile. Pay close attention to the grid spacing and label increment.

- Select "Profiles→Create Profile→Grid" and create major grid lines having the following parameters:

    Horizontal Spacing: 50
    Vertical Spacing: 10
    Height of grid above Datum: 60

- Change the color of the layer "PGRID" by layer to red and the layer name from "PGRID" to "C-PGRID-MAJR". You have successfully created an existing ground profile.

- Select "Profiles→Edit Vertical Alignments". Pick on the "Finished Ground" tab and select the "Center" profile to make the finished ground centerline profile current. Select "OK" to launch the profile editor. Insert a new "PVI" using the toolbar icon having the black dotted triangle with the green center mark. Enter data as identified below. Stations, i.e. "1+00", will have to be entered as absolute numbers, i.e. "100". The "Enter" key will tab between fields.

Station	Elevation	Curve Length
1+00	101.13	
3+25	111.50	
6+50	121.00	
7+75	125.00	50.00
8+75	135.00	
9+74.50	141.50	

- The information that you just entered defines your finished grade alignment. The elevations represent the elevations at each of the stations that you specified. The value for the curve length "50.00" which you entered is the length of the parabolic vertical curve. The horizontal length where the vertical curve meets the finished ground at tangents is 50ft. This is why it is referred to as a 50ft vertical curve. Select "OK" to bring you back to the "Vertical Alignment Editor". Notice that the "Grade Out" represents the calculated grade between the tangent segments and is generated automatically. Close the profile editor. Elect to save the data and import your finished ground profile using the defaults when prompted. Change the color of the layer "PFGC" by layer to blue. Look at it carefully and examine the labeling. Observe the location and information in the vertical curve.

  Another method which users can implement to plot the finished ground vertical alignment on the existing ground profile is through the command "Profiles→FG Vertical Alignments→Import".

- While on the layer "C-ANNO-VPRT" create two 40 scale floating viewports sized 36ft horizontally and 12ft vertically. Display the profile in the upper viewport and the road in the lower viewport. You may need to implement a dview twist to get the entire road in the viewport at the required scale. Don't forget to set your system variables for plotting.

- Do not plot this drawing. You have successfully completed this assignment.

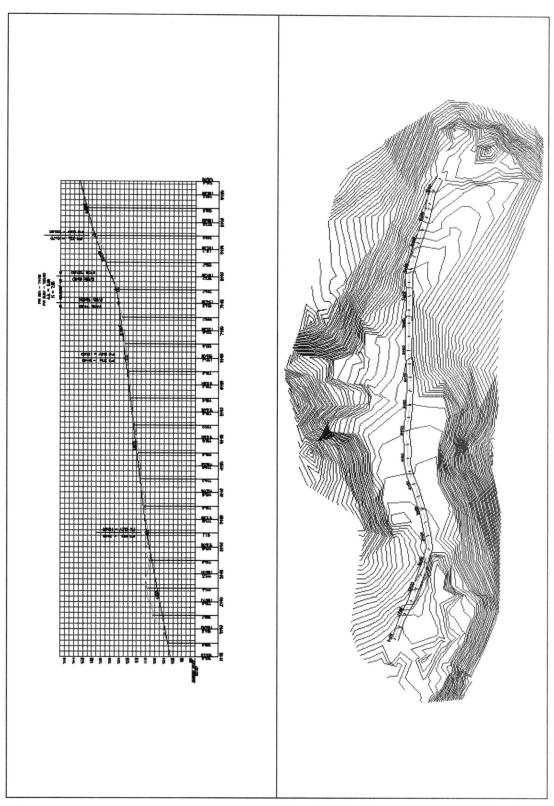

Reference Figure: Assignment 18

NOTES:

Assignment #19
Civil Design Cross Sections

Recommended Assignments Prior to Working this Assignment:

Assignments 1-7, 12-14, 17-18

Required Assignments Prior to Working this Assignment:

Assignment 18

Goals and Objectives

In Assignment 15 we learned that Land Desktop users have the ability to create sections along user defined paths. In this assignment we will explore another way that users can create cross sections through the Civil Design module.

One advantage to creating sections using the Civil Design module is that users do not have to specify the travel paths of each section that they want to process. In the Civil Design Module, users have the ability to sample data along a project alignment at a user specified interval.

The process to create cross sections using the Civil Design module is very similar to the process used when generating profiles.

The steps to create cross sections are identified below:
- Select the desired horizontal alignment.
- Select the desired surface.
- Sample data from the surface which you desire the cross sections to reflect.
- View/Edit your sections.

After this assignment, you will be able to sample surface data at a user specified interval, and use the Civil Design section editor to view your sections. You will also be able to alter the appearance of your cross sections and perform inquiries, such as offsets and elevations for various stations while within the section editor.

Exercise Instructions

- Logon to your workstation and begin a session of Land Desktop. If Land Desktop has been configured on your workstation to display the "Start Up" dialogue box, then cancel this feature so that the AutoCAD model space environment is displayed. If launching Land Desktop brings you directly into the AutoCAD model space environment, then Land Desktop has been configured to begin without the startup dialogue. If the AutoCAD Map "Task Pane" is shown, then close this dialogue box as well. Users that have the ability to customize their profiles can turn these features off permanently by following the procedure described in Assignment #1.

- Use the Land Desktop project manager (Projects→Project Manager) to copy "Assign18-##" to a new project named "Assign19-##" where the "##" reflects the number which was assigned to you on the first day of class. If the number assigned to you is a single digit (for example 4 as opposed to 14), then enter a zero preceding the single digit (such as 04). Refer to Assignment 4 if you can't remember how this was accomplished.

- In the "Copy Project To" "Name:" location type "Assign19". In the "Description:" and "Keywords:" areas, enter the text "Civil Design Cross Sections". Choose "OK" and Land Desktop will ask you if you want to reassociate the copied drawing files so that the copied drawings reference the copied project. Choose "Yes" and then "Close".

- We now have a drawing file named "Assign18-##" in our "Assign19-##" project. It is important that we delete this drawing file before proceeding so that there is no confusion as to which file is the correct "Assign18-##" drawing file. Use Windows Explorer or some other file manager to delete the "Assign18-##.dwg" drawing which resides in your "Assign19-##" project.

- Begin a new drawing by selecting "File→New" to bring up the "New Drawing" dialogue box. Type the "Name:" "Assign19-##" where the "##" reflects the number which was assigned to you on the first day of class.

- We are presently working with Project: "Assign19-##". Make sure that the "Project Name:" displays the correct project, select the "Acad.dwt" template and then "OK" to generate a new drawing in the "DWG" folder of the current project. After selecting the "OK" button, you may be prompted to save changes to the previous drawing session. Since there were no objects in the previous drawing session worth saving, choose not to save changes.

- You will be prompted with the "Load Settings" dialogue box which displays a list of drawing setups to choose from. These setups are saved back to the path determined by the Network Administrator who installed the software. The default path is to the local machine and is displayed above. We will set up our parameters manually, so select "Next" to set up your project with the following parameters:

    Units Area
        Linear Units = Feet
        Angle Units = Degrees
        Angle Display Style = Bearings
        Display Precision Linear = 2
        Display Precision Elevation = 2
        Display Precision Coordinate = 5
        Display Precision Angular = 4
        "Next"

Scale Area
Horizontal = 10
Vertical = 2
Paper Size = 24 x 36(D)
"Next"

Zone Area
"Next"

Orientation Area
"Next"

Text Style Area
Leroy.stp
L100

- Then select "Finish" and a screen will display providing the user with a summary of the settings chosen. Review the settings and select "OK". Although users have the ability to save settings for retrieval with future projects, assignments in this text require that you set up the parameters manually for practice in each project.

- You will not be prompted with the "Create Point Database" dialogue box, because point information was copied from the "Assign18-##" project.

- Create a new layer with the following parameters:

Layer	Color	Linetype
C-ANNO-VPRT	White	Continuous

- While in model space use the external reference command (XR) to attach your Assign18-##.dwg drawing at the coordinates (0,0). Use a scale factor of one and a rotation angle of zero. Zoom extents so that you can view the entire external reference.

    There are several advantages to using the "**External Reference**" feature as opposed to inserting drawing entities when working with sections. Linking these elements using the "External Reference" command helps reduce file size.

    One might ask, why even reference the drawing file if we have access to the project data in an external database? The reason for this is so that we don't bring our section data into a project area that is already occupied with horizontal alignment definitions and/or profile data. As an example, if you created a profile in one drawing linked to Project "A" at the coordinate (5000,5000), then you may not want to import your sections at the coordinate (5000,5000) for Project "A". Instead, you might want to choose a point off in space. One way to ensure that you don't write data to the same database location is to see where data has already been written. The external reference command gives users a visual picture without adding all the linework to your drawing and depending on how external references are set up, the source drawing can be edited simultaneously by other users on a network.

- Select "Alignments→Set Current Alignment→Enter" and choose the centerline alignment that you defined in the previous assignment.

- Change to the Civil Design 2007 menu ("Projects→Workspaces"). Select the menu "Cross Sections→Surfaces→Set Current Surface" and select your "EG" surface and "OK".

- We have now told Land Desktop the path to follow in obtaining our data (alignment selection) and which surface to get our information from. The next step will be to sample sections. Select "Cross Sections→Existing Ground→Sample from Surface". The **"Swath widths"** tell Land Desktop how far to sample data on each side of our alignment. Set the left and right swath widths to 50ft. The "Tangents", "Curves" and "Spirals" define the interval that we want to sample. Set these values to 25ft. In the area for additional sample control, choose to sample the alignment "Start" and "End" only. When you are finished, your dialogue box will appear as follows:

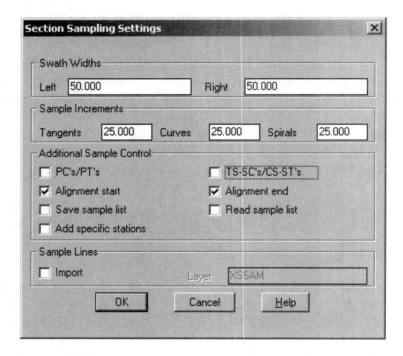

Select "OK" to accept and exit the "Section Sampling Settings". Accept the default start and end sampling stations which will correspond to the start and ends of your alignment.

- View your sections using the command "Cross Sections→View/Edit Sections". Take a look at your command line. The prompt is set to view the next section. If we press the "Enter" key Land Desktop shows us the next section. Toggle through a few of your sections and take a look at them.

- Use the "View" option while in the "View/Edit Sections" command by typing "V" to turn on all of the options with the exception of the "Point codes". Notice that you also have the option to modify the vertical exaggeration in this dialogue box. Now use the "Next" and "Previous" options to run back and fourth through your sections while observing them.

- Now use the "Zoom" option to zoom in and out of your cross sections and observe what happens if you zoom in and proceed to the next cross section.

- Try using the "ID" command to identify points in your cross section and observe the sign of the offset if you inquire offsets to the left of the alignment. Exit the "Section View Editor".

- Export your section point data to an external text file using the command "Cross Sections→Point Output→Tplate Points to File". Save the file using the name Tplate.txt to the Survey directory of the corresponding project. Elect to only export existing ground data for the full alignment length.
- Launch a session of notepad to view the text file that you just created.

- While on the layer "C-ANNO-VPRT" create two 40 scale floating viewports sized 36ft horizontally and 12ft vertically. Display the profile in the upper viewport and the road in the lower viewport. You may need to implement a dview twist to get the entire road in the viewport at the required scale. Don't forget to set your system variables for plotting.

- The figure will look identical to the figure in Assignment 18 and is shown at the end of this assignment for reference. Do not plot this drawing. You have successfully completed this assignment.

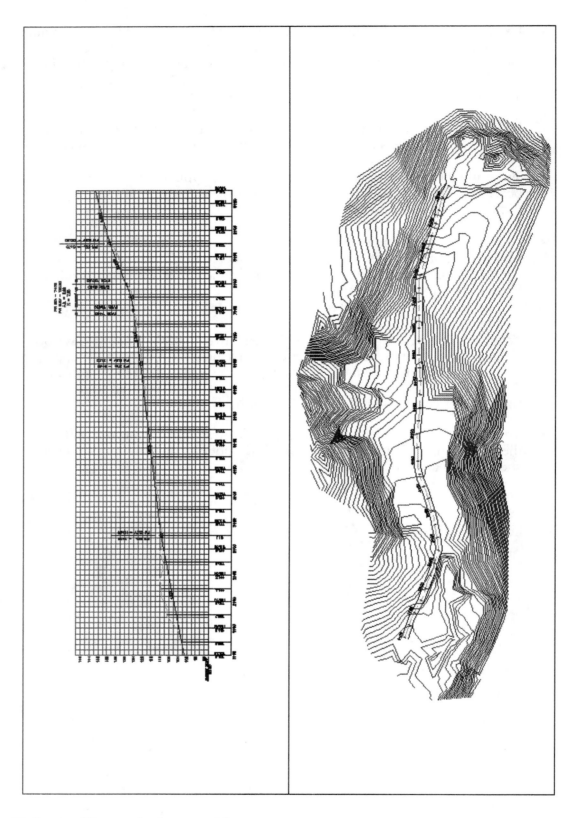

Reference Figure: Assignment 19

Assignment #20
Civil Design Section Plotting

Recommended Assignments Prior to Working this Assignment:

Assignments 1-7, 12-14, 16, 17-19

Required Assignments Prior to Working this Assignment:

Assignments 18-19

Goals and Objectives

In Assignment 19 we learned how to create cross sections using the Civil Design module and how to view the sections that we created using the "View/Edit Sections" routine. One of the advantages to using the "View/Edit Sections" routine is that a user can rapidly toggle through multiple sections and get a feel for what is happening. Users might then elect to edit errors in their surface and generate new sections before continuing. In addition, there are a number of utilities available in the section editor which will allow users to zoom and identify approximate offsets/elevations by simply picking points on the screen while viewing sections.

You may have discovered that AutoCAD object snaps do not work while in the view/edit sections routine. Since entities have not been inserted into the drawing environment, there are no objects for users to snap to. In this assignment, we will learn how to bring the section data into the AutoCAD environment and use some of Land Desktop's "Section Utilities" to make inquiries about the sections that we import.

Exercise Instructions

- Logon to your workstation and begin a session of Land Desktop. If Land Desktop has been configured on your workstation to display the "Start Up" dialogue box, then cancel this feature so that the AutoCAD model space environment is displayed. If launching Land Desktop brings you directly into the AutoCAD model space environment, then Land Desktop has been configured to begin without the startup dialogue. If the AutoCAD Map "Task Pane" is shown, then close this dialogue box as well. Users that have the ability to customize their profiles can turn these features off permanently by following the procedure described in Assignment #1.

- Use the Land Desktop project manager (Projects→Project Manager) to copy "Assign19-##" to a new project named "Assign20-##" where the "##" reflects the number which was assigned to you on the first night of class. If the number assigned to you is a single digit (for example 4 as opposed to 14), then enter a zero preceding the single digit (such as 04). Refer to Assignment 4 if you can't remember how this was accomplished.

- In the "Copy Project To" "Name:" location type "Assign20-##". In the "Description:" and "Keywords:" areas, enter the text "Civil Design Section Plotting". Choose "OK" and Land Desktop will ask you if you want to reassociate the copied drawing files so that the copied drawings reference the copied project. Choose "Yes" and then "Close".

- We now have a drawing file named "Assign19-##" in our "Assign20-##" project. It is important that we delete this file before proceeding so that there is no confusion as to which file is the correct "Assign19-##" drawing file. Use Windows Explorer or some other file manager to delete the "Assign19-##.dwg" drawing which resides in your "Assign20-##" project.

- Begin a new drawing by selecting "File→New" to bring up the "New Drawing" dialogue box. Type the "Name:" "Assign20-##" where the "##" reflects the number which was assigned to you on the first day of class.

- We are presently working with Project: "Assign20-##". Make sure that the "Project Name:" displays the correct project, select the "Acad.dwt" template and then "OK" to generate a new drawing in the "DWG" folder of the current project. After selecting the "OK" button, you may be prompted to save changes to the previous drawing session. Since there were no objects in the previous drawing session worth saving, choose not to save changes.

- You will be prompted with the "Load Settings" dialogue box which displays a list of drawing setups to choose from. These setups are saved back to the path determined by the Network Administrator who installed the software. The default path is to the local machine and is displayed above. We will set up our parameters manually, so select "Next" to set up your project with the following parameters:

    Units Area
        Linear Units = Feet
        Angle Units = Degrees
        Angle Display Style = Bearings
        Display Precision Linear = 2
        Display Precision Elevation = 2
        Display Precision Coordinate = 5
        Display Precision Angular = 4
        "Next"

    Scale Area
        Horizontal = 10
        Vertical = 2
        Paper Size = 24 x 36(D)
        "Next"

> Zone Area
> "Next"
>
> Orientation Area
> "Next"
>
> Text Style Area
> Leroy.stp
> L100

- Then select "Finish" and a screen will display providing the user with a summary of the settings chosen. Review the settings and select "OK". Although users have the ability to save settings for retrieval with future projects, assignments in this text require that you set up the parameters manually for practice in each project.

- You will not be prompted with the "Create Point Database" dialogue box, because point information was copied from the "Assign19-##" project.

- While in model space use the external reference command (XR) to attach your Assign18-##.dwg drawing at the coordinates (0,0). Use a scale factor of one and a rotation angle of zero. Zoom extents so that you can view the entire external reference.

- Change to the Civil Design Module and select "Alignments→Set Current Alignment→Enter" and choose the centerline alignment that you defined in a previous assignment.

- View your sections using the command "Cross Sections→View/Edit Sections". These sections should look familiar to you. View a few of the sections and exit the command.

- Create a few new layers with the following parameters:

Name	Color	Linetype
C-ANNO-VPRT	White	Continuous
XEG	Green	Dashed2
XGRID	245	Continuous

- Select "Cross Sections→Section Plot→Settings". Toggle on everything except the right of way lines and template. We will address these in future lessons. Select "Section Layout" and fill out the dialogue box as follows:

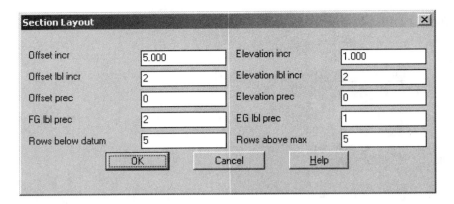

The increment is the distance between grid lines.
The label increment is the nth grid in the direction being labeled.
The precision is the number of decimal places to display on the labels.
The label precision is the number of places to display for centerline elevations.
The rows below the datum and above the max dictate additional grid below and above the section data.

- Select the "Page Layout" option and fill out the dialogue box as follows:

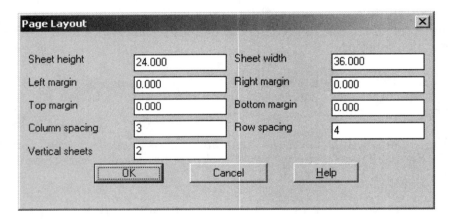

The actual AutoCAD vertical elements are the product of the spacing shown in the table above and the vertical distortion between the horizontal and vertical cross section scales.

- Select "Cross Sections→Section Plot→All". Select the defaults for the beginning and ending stations of your alignment and insert your sections at the coordinate (18000,5000).

- Use the command "Cross Sections→Section Utilities→Select by Station" to select Station 500. Now use the command "Cross Sections→Section Utilities→List Offset/Elevation" to list offsets and elevations for the current section. Notice that you have the ability to use your object snaps and observe what happens if you inquire offsets for an area in a section that is not current.

- Use the command "Cross Sections→Section Utilities→Select by Point" to pick a point inside the section for station 400. Now use the command "Cross Sections→Section Utilities→List Slope/Grade" to inquire slopes and grades for the current section. Note: Two points must be selected to yield slopes and grades. Observe the result and determine the units for slopes and grades.

- Use the command "Cross Sections→Section Utilities→Zoom to Station" and type in station 300. Use 100 when prompted for a zoom factor. Use the command "Cross Sections→Section Utilities→List Area" to compute the area of one grid cell. Now use the AutoCAD "Area" command to list the area of one grid cell and compare the areas.

Since Civil Design cross sections know where they were initially inserted into the CAD environment, the section utilities may not work appropriately if cross sections are moved after to being plotted to the drawing. Therefore, users should make an effort to place cross sections in an area where they are not anticipated to move.

This often creates a problem when cross sections are plotted to the drawing more than one time. The command "Cross Sections→Section Plot→Undefine Section" will delete cross section definitions and should be run prior to subsequently plotting cross sections to the drawing environment.

- Zoom extents and while on the layer "C-ANNO-VPRT" create a 10 scale landscape floating viewport which displays the first 3 cross sections generated in this assignment.

- Save and exit your drawing. You have successfully completed this assignment.

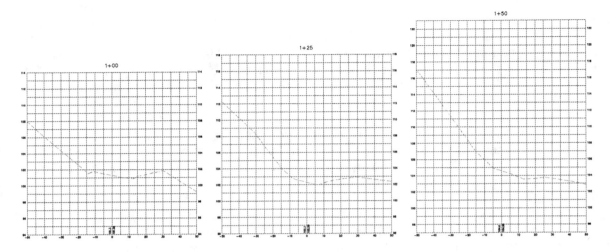

Reference Figure: Assignment 20

Assignment #21
Drawing Civil Design Templates

Recommended Assignments Prior to Working this Assignment:

Assignments 1-7, 12-14, 17-19

Required Assignments Prior to Working this Assignment:

Assignments 18-19

Goals and Objectives

    A **"Template"** is a typical cross section containing finished ground surface data. Templates generally consist of two surface types. The first surface type is known as **"Normal"**. This surface type usually contains elements in the upper (highest elevations) portion of the template. Asphalt and/or subassemblies (i.e. concrete curb and gutter) would be considered normal surfaces. The second surface type contains elements below this. This surface type is known as the "subgrade" surface type. A **"subgrade"** surface type might contain structural elements such as class II base rock laid underneath the

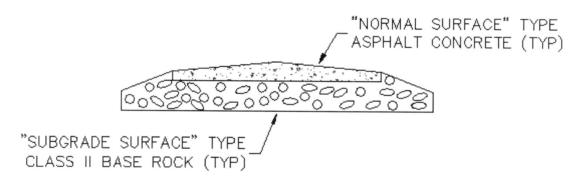

asphalt course. Although a template might contain one or both of these surface types, we will explore normal surface templates in the coming assignments.

    You will find, in your experience as a Civil/Survey technician, that sheets of detailed cross sections are seldom included with improvement drawing sets. Instead, the improvement drawings generally show a typical finished ground roadway structural section. Elevations in these cases are called out on the grading plan, which is a 2 dimensional horizontal view of the area.

    Although not always required with drawings, cross sections give us a good feel for what is happening in the vertical direction, so they are very handy throughout the design process, but it might not always be worth your while to create detailed cross sections with both surface types.

    Another advantage to creating cross sections is that they provide a method for computing earthwork quantities (cut and fill) for your project. Users need to keep in mind that earthwork computations are calculated from the datum (bottom) of a template.

A template defined as only having a "Normal" surface can generate erroneous earthwork results if the "Normal" surface is not set up to simulate base rock, and subassemblies (Curb, Gutter & Sidewalk, etc).

If you desire to create only "Normal Surface" templates, you desire not to create subassemblies, and you want your earthwork computations to yield accurate results, then you may choose to include subassembly areas in the "Normal Surface" template. In the preliminary stages of roadway design, a user saves time proceeding in this manner, because they do not have to define multiple subassemblies or set up the section control for them.

Before drawing templates, users should set the template path. Land Desktop by default writes defined templates to its data directory for the local machine. This means that you could conceivably have templates from 500+ projects in one location. The consequences can be good or bad depending on template organization. Furthermore, if someone were to open a project on another network workstation, then they might not have access to the same project templates if Land Desktop were not set up appropriately. Since all of our other data is project specific, why not make your templates project specific as well? This manual recommends that you create a directory in your project called Tplate and set this as the active template path.

There are several different methods which may be implemented when creating templates. Templates may be drawn using the "Draw Template" routine or by simply drawing a 2D polyline. In any event, there are some rules that need to be followed:

Templates must be drawn to their appropriate scale.

The **"Draw Template"** command takes in to account the vertical exaggeration. However, if users wish to draw a template using the 2D polyline routine, then they must manually include the vertical exaggeration when it is drawn.

Templates should be drawn counterclockwise from right to left.

If a template is symmetrical, then only one side needs to be drawn.

After this lesson, users will be able to draw templates with the "Draw Template" routine and with a 2D polyline. We will see after this lesson that only half of a symmetrical template needs to be created and that, when creating templates from a 2D polyline, the vertical exaggeration needs to be manually accounted for.

Exercise Instructions

- Logon to your workstation and begin a session of Land Desktop. If Land Desktop has been configured on your workstation to display the "Start Up" dialogue box, then cancel this feature so that the AutoCAD model space environment is displayed. If launching Land Desktop brings you directly into the AutoCAD model space environment, then Land Desktop has been configured to begin without the startup dialogue. If the AutoCAD Map "Task Pane" is shown, then close this dialogue box as well. Users that have the ability to customize their profiles can turn these features off permanently by following the procedure described in Assignment #1.

- Use the Land Desktop project manager (Projects→Project Manager) to copy "Assign19-##" to a new project named "Assign21-##" where the "##" reflects the number which was assigned to you on the first day of class. If the number assigned to you is a single digit (for example 4 as opposed to 14), then enter a zero preceding the single digit (such as 04). Refer to Assignment 4 if you can't remember how this was accomplished.

- In the "Copy Project To" "Name:" location type "Assign21-##" where the "##" reflects the number which was assigned to you on the first day of class. In the "Description:" and "Keywords:" areas, enter the text "Drawing Civil Design Templates". Choose "OK" and Land Desktop will ask you if you want to reassociate the copied drawing files so that the copied drawings reference the copied project. Choose "Yes" and then "Close".

- We now have a drawing file named "Assign19-##" in our "Assign21-##" project. It is important that we delete this file before proceeding so that there is no confusion as to which file is the correct "Assign19-##" drawing file. Use Windows Explorer or some other file manager to delete the "Assign19-##.dwg" drawing which resides in your "Assign21-##" project.

- Begin a new drawing by selecting "File→New" to bring up the "New Drawing" dialogue box. Type the "Name:" "Assign21-##" where the "##" reflects the number which was assigned to you on the first day of class.

- We are presently working with Project: "Assign21-##". Make sure that the "Project Name:" displays the correct project, select the "Acad.dwt" template and then "OK" to generate a new drawing in the "DWG" folder of the current project. After selecting the "OK" button, you may be prompted to save changes to the previous drawing session. Since there were no objects in the previous drawing session worth saving, choose not to save changes.

- You will be prompted with the "Load Settings" dialogue box which displays a list of drawing setups to choose from. These setups are saved back to the path determined by the Network Administrator who installed the software. The default path is to the local machine and is displayed above. We will set up our parameters manually, so select "Next" to set up your project with the following parameters:

    Units Area
        Linear Units = Feet
        Angle Units = Degrees
        Angle Display Style = Bearings
        Display Precision Linear = 2
        Display Precision Elevation = 2
        Display Precision Coordinate = 5
        Display Precision Angular = 4
        "Next"

Scale Area
    Horizontal = 20
    Vertical = 5
    Paper Size = 8 x 11(A)
    "Next"

Zone Area
    "Next"

Orientation Area
    "Next"

Text Style Area
    Leroy.stp
    L100

- Then select "Finish" and a screen will display providing the user with a summary of the settings chosen. Review the settings and select "OK". Although users have the ability to save settings for retrieval with future projects, assignments in this text require that you set up the parameters manually for practice in each project.

- You will not be prompted with the "Create Point Database" dialogue box, because point information was copied from the "Assign19-##" project.

- While in model space use the external reference command (XR) to attach your Assign19-##.dwg drawing at the coordinates (0,0). Use a scale factor of one and a rotation angle of zero. Zoom extents so that you can view the entire external reference.

- Create a few new layers with the following parameters:

Name	Color	Linetype
C-ANNO-VPRT	White	Continuous
C-TPLT-FGND	Green	Continuous

- Using Windows Explorer or some other file management utility, create a new folder in your current project called "Tplate".

- While in the Civil Design module, use the command "Cross Sections→Set Template Path" to set the path to the "Tplate" directory that you just created. You will have to toggle off the check box in the current dialogue box to set the new path. This path will be used in the next assignment when template definition is written to an external file using the define template routine.

- Your template will not take up very much space, so zoom in to an area on your screen about the size of a dime. While on the layer "C-TPLT-FGND" select the command "Cross Sections→Draw Template". Proceed to draw your template by first selecting an arbitrary point on your screen and then selecting the grade option. Make sure that you pay close attention to what is happening. Watch your command line. For the grade enter "-2" (for 2%). The corresponding offset for this grade will be "-10" (for 10ft to the left). Now select the relative option. Use "0" for the change in offset and enter "-1" for the change in elevation. (We will not define a subgrade surface type but this 1ft will simulate our base rock for a future volume computation.) Using the grade option, enter a grade of 2% and an offset of 10ft. Use the "C" option to close your polyline and press "Enter" one more time to terminate the command.

Notice the vertical exaggeration of the cross section. Make sure that you understand the difference between specifying grade and specifying relative. Repeat the draw template command and draw the template described above several times. Continue to draw the template until you can draw the template in less than 30seconds.

- Now we will use the 2D polyline command to draw the same template. Type "PL" to execute the polyline command. Choose an arbitrary point on your screen somewhere beside the templates that you just drew and proceed to draw the template as follows:

    @-10,-.2
    @0,-1
    @10,.2
    C for close.

    Why do you suppose the two templates do not look the same?

- Let's try to draw this template again using the 2D polyline method. Type "PL" to execute the polyline command. Choose an arbitrary point on your screen somewhere beside the last template that you just drew and proceed to draw the template as follows:

    @-10,-.8
    @0,-4
    @10,.8
    C for close.

    What is the relationship between the first and second set of numbers for the two polyline templates?

- While on the layer "C-ANNO-VPRT" create a 20 scale viewport on an 8 ½"× 11" sheet of paper in plan view. The viewport is to have the landscape orientation and the lower left hand corner of the viewport must be at the origin (0,0). The viewport should only show the templates that you drew in this assignment. If you drew 10 templates using the command "Cross Sections→Draw Template", then show all 10 templates using this command plus the two templates drawn using the polyline technique.

- Save and exit your drawing. You have successfully completed this assignment.

Introduction to Land Desktop 2007

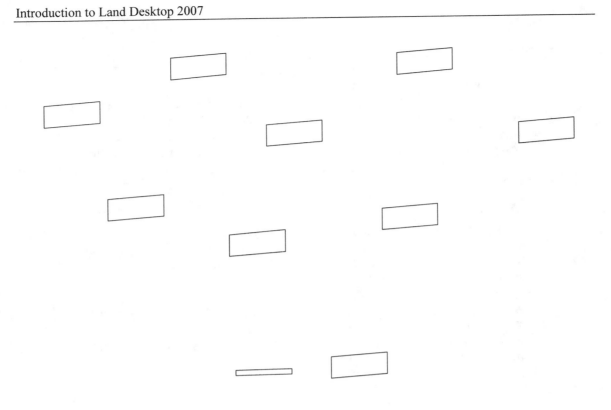

Reference Figure: Assignment 21

Assignment #22
Defining Civil Design Templates

Recommended Assignments Prior to Working this Assignment:

Assignments 1-7, 12-14, 17-19, 21

Required Assignments Prior to Working this Assignment:

Assignments 18-19

Goals and Objectives

In our last lesson, we learned how to draw templates. We also learned how to set the path that Land Desktop uses when retrieving templates and saving template modifications. We learned that Land Desktop doesn't actually write the template to an external database until the template is defined.

When a user elects to define a template, there are several pieces of information that Land Desktop requests. The first bit of information that Land Desktop wants is the **"Finished Ground Reference Point"**. Land Desktop uses the finished ground reference point as a guide for which your template tracks your primary alignment. This reference point will most often correspond to the horizontal alignment that we defined as the centerline of road. This is the same alignment that we used when we sampled existing ground from our TIN. The finished ground reference point will most often be the starting point of your template. Land Desktop will next prompt the user to identify the template as **"Symmetrical"** or **"Asymmetrical"**. Users will be asked whether the surface type is **"Normal"** or **"Subgrade"** (See Assignment 21 for definitions of Normal and Subgrade) and will be prompted to enter a **Name**. After entering a name the **"Connection Point Out"** must be specified. This point is the point that Land Desktop will use to connect subassemblies such as curb, gutter and sidewalk. Subassemblies do not have to be defined to set the connection point out. These points will most often be the left and right extremes of the template. The **"Datum"** will have to be specified next. The datum is the bottom of our template. Since everything above the datum represents our surfaces, (e.g., asphalt, concrete and base, etc.) the datum establishes the boundary between our surfaces and the earth. Defining the datum will later allow us to perform earthwork computations. The datum should always be defined from left to right and include everything except the top surface. Subassemblies are the last thing that users will be prompted for when defining a template. Subassemblies do not have to be chosen to finish your template definition.

Now that we have learned how to define the template, the next step will be to set the template up so that we can later attach the template to our alignments and profiles using design control. We will use the "Edit Template" routine to add surface control to our defined template. One type of surface control is a transition region. **"Transition Regions"** are used to stretch templates in the horizontal and vertical directions to account for variations in lane widths due to features as turning lanes. Transition regions give us the ability to use a single template for a project to accommodate varying conditions. Eight transition regions can be set for each side (Left and Right) of a template. A Transition region consists of two points (Control and Region). The **"Control Point"** represents the point which our alignments and profiles are attached. The **"Region Point"** is a point between the control point and template centerline representing the furthest edge of the template that will stretch when the control point moves along the transition lines. By definition, a control point can never be closer to the centerline than a region point. When we define a template's **"Surface Transition"**, Land Desktop asks the user whether the connection is **"Pinned"** or **"Dynamic"**. For pinned surface transitions, the template's FG reference point is pinned in place (usually to the centerline alignment). For the dynamic surface transition, the templates FG reference point is free to move as the control point expands from roadway widening. The "Pinned" option is the most common and should be used whenever the user has any doubts about which method is going to yield more pleasing results. Land Desktop next asks for the **"Transition Region Type"**. Transition regions are either **"Constrained"** or **"Free"**. Constrained transition regions can only be stretched away from the template centerline while free transition regions may be stretched in both directions. The "Free" option is the most common and should be used whenever the user has any doubts about which method to employ. Users also have the ability to hold **"Grade"** or **"Elevation"** when setting up their transitions. The grade option will maintain the template surface grade when horizontal transition lines create roadway widening. If the grade option is selected, a reference point must also be supplied. The reference point represents the grade that is to be held while the template is being stretched. The Elevation option will maintain a constant elevation triggered by the template outer edge elevation and extending horizontally to the corresponding horizontal transition line. Users should keep in mind that one of these options must be selected, but neither of them is used if or when vertical transitioning is applied in "Design control".

Exercise Instructions

- Logon to your workstation and begin a session of Land Desktop. If Land Desktop has been configured on your workstation to display the "Start Up" dialogue box, then cancel this feature so that the AutoCAD model space environment is displayed. If launching Land Desktop brings you directly into the AutoCAD model space environment, then Land Desktop has been configured to begin without the startup dialogue. If the AutoCAD Map "Task Pane" is shown, then close this dialogue box as well. Users that have the ability to customize their profiles can turn these features off permanently by following the procedure described in Assignment #1.

- Use the Land Desktop project manager (Projects→Project Manager) to copy "Assign19-##" to a new project named "Assign22-##" where the "##" reflects the number which was assigned to you on the first day of class. If the number assigned to you is a single digit (for example 4 as opposed to 14), then enter a zero preceding the single digit (such as 04). Refer to Assignment 4 if you can't remember how this was accomplished.

- In the "Copy Project To" "Name:" location type "Assign22-##" where the "##" reflects the number which was assigned to you on the first day of class. In the "Description:" and "Keywords:" areas, enter the text "Defining Civil Design Templates". Choose "OK" and Land Desktop will ask you if you want to reassociate the copied drawing files so that the copied drawings reference the copied project. Choose "Yes" and then "Close".

- We now have a drawing file named "Assign19-##" in our "Assign22-##" project. It is important that we delete this file before proceeding so that there is no confusion as to which file is the correct "Assign19-##" drawing file. Use Windows Explorer or some other file manager to delete the "Assign19-##.dwg" drawing which resides in your "Assign22-##" project.

- Begin a new drawing by selecting "File→New" to bring up the "New Drawing" dialogue box. Type the "Name:" "Assign22-##" where the "##" reflects the number which was assigned to you on the first day of class.

- We are presently working with Project: "Assign22-##". Make sure that the "Project Name:" displays the correct project, select the "Acad.dwt" template and then "OK" to generate a new drawing in the "DWG" folder of the current project. After selecting the "OK" button, you may be prompted to save changes to the previous drawing session. Since there were no objects in the previous drawing session worth saving, choose not to save changes.

- You will be prompted with the "Load Settings" dialogue box which displays a list of drawing setups to choose from. These setups are saved back to the path determined by the Network Administrator who installed the software. The default path is to the local machine and is displayed above. We will set up our parameters manually, so select "Next" to set up your project with the following parameters:

    Units Area
        Linear Units = Feet
        Angle Units = Degrees
        Angle Display Style = Bearings
        Display Precision Linear = 2
        Display Precision Elevation = 2
        Display Precision Coordinate = 5
        Display Precision Angular = 4
        "Next"

    Scale Area
        Horizontal = 20
        Vertical = 5
        Paper Size = 8 x 11(A)
        "Next"

    Zone Area
        "Next"

    Orientation Area
        "Next"

22.3

Text Style Area
Leroy.stp
L100

- Then select "Finish" and a screen will display providing the user with a summary of the settings chosen. Review the settings and select "OK". Although users have the ability to save settings for retrieval with future projects, assignments in this text require that you set up the parameters manually for practice in each project.

- You will not be prompted with the "Create Point Database" dialogue box, because point information was copied from the "Assign19-##" project.

- While in model space use the external reference command (XR) to attach your Assign19-##.dwg drawing at the coordinates (0,0). Use a scale factor of one and a rotation angle of zero. Zoom extents so that you can view the entire external reference.

- Create a few new layers with the following parameters:

Name	Color	Linetype
C-ANNO-VPRT	White	Continuous
C-TPLT-FGND	Green	Continuous

- Using Windows Explorer, create a new folder in your current project called "Tplate". This path will be used when template definition is written to an external file using the define template routine later in the lesson.

- While in the Civil Design module, use the command "Cross Sections→Set Template Path" to set the path to the "Tplate" directory that you just created. You will have to toggle off the check box in the current dialogue box to set the new path.

- Your template will not take up very much space so zoom in to an area on your screen about the size of a dime. While on the layer "C-TPLT-FGND" select the command "Cross Sections→Draw Template". Proceed to draw your template by first selecting an arbitrary point on your screen, and then selecting the grade option. Make sure that you pay close attention to what is happening. Watch your command line. For the grade enter -2%. The corresponding offset for this grade will be -10ft. Now select the relative option. Use 0ft for the change in offset and enter -1ft for the change in elevation. (We will not define a subgrade surface type but this 1ft will simulate our base rock for a future volume computation.) Using the grade option, enter a grade of 2% and an offset of 10ft. Do not close your template as in the last lesson. Instead, "Enter" twice will terminate the command.

- Now that we have drawn our template, we need to define the template. Use the command "Cross Sections→Templates→Define Template". For the "Finished Ground Reference Point" use your object snap to select the point where you began drawing your template. Choose "Yes" when asked whether or not the template is symmetrical and when prompted, select the template. We learned from our last assignment that the surface type for this template will be "Normal". Enter a description of "Road". Make sure that you use your object snaps for the next series of questions. The "Connection Point Out" will be the upper left hand corner of the template. There will only be one "Datum". The datum will begin with the upper left hand corner and work around the template from left to right in the counterclockwise direction. Accept the defaults for the "Null" subassembly attachments. Choose to save your template with the name "Road". Repeat this process several times until you are able to define your template without looking at the instructions.

- We are now finished with the template entities on our screen, so erase them from your drawing.

- The next step will be to set up the surface control. Use the command "Cross Sections→Templates→Edit Template". Select the template that you just created. The insertion point will be about the FG reference point that you specified when you defined your template. Notice that the template now has symmetry. You should also notice that the template comes in on the current layer. The datum shows up as a dotted line in a different color. The datum is not an AutoCAD entity and will disappear when AutoCAD's video is regenerated. Type "SR" to modify the surface control. Now type "TR" to modify the transition settings. First choose to set up the left by typing "L". Choose region "1". The first left transition region point will be the upper left hand corner of the template. Select the options "Pinned, Free, Same & Grade" and supply the FG reference point as the reference point for the grade option. Make sure that you understand the options that you selected. Type "X" to back up to the previous menu and choose to set up the right the same way that you set up the left. Now exit three times and choose to "Save" your template with the same name that you gave it before. After saving "X" will get you out of the command. Repeat this process several times until you are able to set up surface control for your template without looking at the instructions.

- We will next use the edit routine to check the template definition. Select "Cross Sections→Templates→Edit Template" and once again insert the template into your drawing. Make sure that your datum looks correct. Type "SR" and then "TR" and notice that Land Desktop designates the left and right transitions with an "L1" and an "R1". Exit one time and choose the "Connect" option to view the connections. Exit the "Edit Template" routine.

- Zoom extents and while on the layer "C-ANNO-VPRT" create a 20 scale viewport on an 8 ½″ × 11″ sheet of paper in plan view. The viewport is to have the landscape orientation and the lower left hand corner of the viewport must be at the origin (0,0). The viewport should only show the templates that you drew in this assignment.

- Save and exit your drawing. You have successfully completed this assignment.

Introduction to Land Desktop 2007

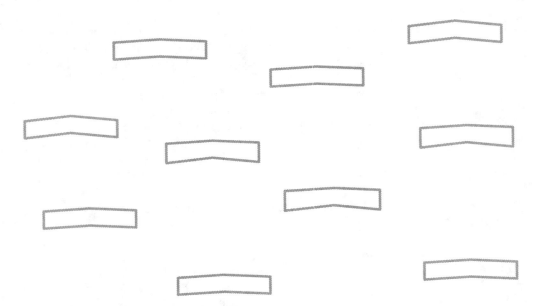

Reference Figure: Assignment 22

Assignment #23
Civil Design Control-Templates

Recommended Assignments Prior to Working this Assignment:

Assignments 1-7, 12-14, 17-22

Required Assignments Prior to Working this Assignment:

Assignments 18-19, 22

Goals and Objectives

In the last couple of lessons we learned how to draw and define templates. We then learned how to set up surface control for our templates so that we could later attach templates to our alignments and profiles. In this lesson, we will take the next step in the continuation of our template studies with an introduction to **"Design Control"**. Design Control allows us to specify the templates that we wish to use. It also gives users the ability to attach alignments and profiles, specify slopes and other design criteria.

After this assignment, you will be able to specify templates to be used in design control and establish cut and fill slopes for the finished ground alignment. We will address attaching alignments and profiles in future lessons.

Exercise Instructions

- Logon to your workstation and begin a session of Land Desktop. If Land Desktop has been configured on your workstation to display the "Start Up" dialogue box, then cancel this feature so that the AutoCAD model space environment is displayed. If launching Land Desktop brings you directly into the AutoCAD model space environment, then Land Desktop has been configured to begin without the startup dialogue. If the AutoCAD Map "Task Pane" is shown, then close this dialogue box as well. Users that have the ability to customize their profiles can turn these features off permanently by following the procedure described in Assignment #1.

- Use the Land Desktop project manager (Projects→Project Manager) to copy "Assign22-##" to a new project named "Assign23-##" where the "##" reflects the number which was assigned to you on the first day of class. If the number assigned to you is a single digit (for example 4 as opposed to 14), then enter a zero preceding the single digit (such as 04). Refer to Assignment 4 if you can't remember how this was accomplished.

- In the "Copy Project To" "Name:" location type "Assign23-##" where the "##" reflects the number which was assigned to you on the first day of class. In the "Description:" and "Keywords:" areas, enter the text "Civil Design

Control - Templates". Choose "OK" and Land Desktop will ask you if you want to reassociate the copied drawing files so that the copied drawings reference the copied project. Choose "Yes" and then "Close".

- We now have a drawing file named "Assign22-##" in our "Assign23-##" project. It is important that we delete this file before proceeding so that there is no confusion as to which file is the correct "Assign22-##" drawing file. Use Windows Explorer or some other file manager to delete the "Assign22-##.dwg" drawing which resides in your "Assign23-##" project.

- Begin a new drawing by selecting "File→New" to bring up the "New Drawing" dialogue box. Type the "Name:" "Assign23-##" where the "##" reflects the number which was assigned to you on the first day of class.

- We are presently working with Project: "Assign23-##". Make sure that the "Project Name:" displays the correct project, select the "Acad.dwt" template and then "OK" to generate a new drawing in the "DWG" folder of the current project. After selecting the "OK" button, you may be prompted to save changes to the previous drawing session. Since there were no objects in the previous drawing session worth saving, choose not to save changes.

- You will be prompted with the "Load Settings" dialogue box which displays a list of drawing setups to choose from. These setups are saved back to the path determined by the Network Administrator who installed the software. The default path is to the local machine and is displayed above. We will set up our parameters manually, so select "Next" to set up your project with the following parameters:

    Units Area
        Linear Units = Feet
        Angle Units = Degrees
        Angle Display Style = Bearings
        Display Precision Linear = 2
        Display Precision Elevation = 2
        Display Precision Coordinate = 5
        Display Precision Angular = 4
        "Next"

    Scale Area
        Horizontal = 40
        Vertical = 10
        Paper Size = 24 x 36(D)
        "Next"

    Zone Area
        "Next"

<u>Orientation Area</u>
"Next"

<u>Text Style Area</u>
Leroy.stp
L100

- Then select "Finish" and a screen will display providing the user with a summary of the settings chosen. Review the settings and select "OK". Although users have the ability to save settings for retrieval with future projects, assignments in this text require that you set up the parameters manually for practice in each project.

- You will not be prompted with the "Create Point Database" dialogue box, because point information was copied from the "Assign23-##" project.

- While in model space use the "Insert→Block→Browse" command (I) to insert your Assign18-##.dwg drawing at the coordinates (0,0). Use a scale factor of one and a rotation angle of zero. Zoom extents so that you can view the entire block and explode the block one time.

- Create a new layer having the following parameters:

<u>Name</u>	<u>Color</u>	<u>Linetype</u>
XEG	Green	Dashed
XGRID	245	Continuous
XFG	Blue	Continuous

- While in the Civil Design module, use the command "Cross Sections→Set Template Path" to set the path to the "Tplate" directory for the current project.

- In the last assignment, we created a symmetrical template with a 2% cross slope and 10ft half width. In this lesson, create another symmetrical template with a 2% cross slope and 15ft half width. Name the template "Road15". Define and set up the surface control the same way that you did for the 10ft template. See the previous assignment if you can't remember how this was accomplished.

- Set your centerline alignment current. Select "Cross Sections→View/Edit Sections" to view your sections. Take a quick look at a few of them. Notice that the sections only show the existing ground and the background grid. The finished ground is not shown even though templates have been drawn.

- Specifying the templates used along our alignment will be our next task. Select "Cross Sections→Design Control→Edit Design Control". The routine will prompt you for a start station and end station. The default will be the start and end of your alignment. Accept 100 as the start and specify 300 as the end.

- Next choose "Template Control" and select the "Road" template that you defined in the last lesson. Refer to the graphic below for reference.

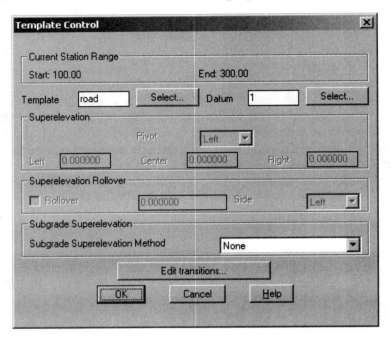

- Select "OK" to allow Land Desktop to process your sections. If you have any errors, ignore them for the time being by selecting "OK". You have successfully bound the "Road" template to your centerline alignment about the FG reference point for the station range 100-300.

- Select "Cross Sections→View/Edit Sections" to view your sections. Use the "View" option to toggle off the "Top Surface" and "ROW Lines" and to change the "Vertical Factor" to 4. Notice that the template is shown for the first 300ft of alignment and then the finished ground disappears.

- Select "Cross Sections→Design Control→Edit Design Control". The routine will once again prompt you for a start station and end station. The default will be the start and end of your alignment. Type 325 for the start and 600 for the end and select "OK". Choose "Template Control" and select the "Road15" template. Select "OK" to bring you back to the "Template Control" dialogue box, "OK" to bring you back to the "Design Control" dialogue box and "OK" again to allow Land Desktop to process your sections. If you have any errors, ignore them for the time being by selecting "OK". You have successfully bound the "Road15" template to your centerline alignment about the FG reference point for the station range 325-600.

- Select "Cross Sections→Design Control→Edit Design Control". The routine will once again prompt you for a start station and end station. The default will be the start and end of your alignment. Type 625 for the start and accept the default end station. Choose "Template Control" and select the "Road" template. Select "OK" to bring you back to the "Template Control" dialogue box and "OK" again to bring you back to the "Design Control" dialogue box. Choose the "Slopes" option and enter a 2 (This means an elevation of 1ft per

2ft of horizontal) for all of the typical cuts and fills. Use the "Simple" type for cuts and fills. Leave the maximum slope at 0. Refer to the graphic below for reference.

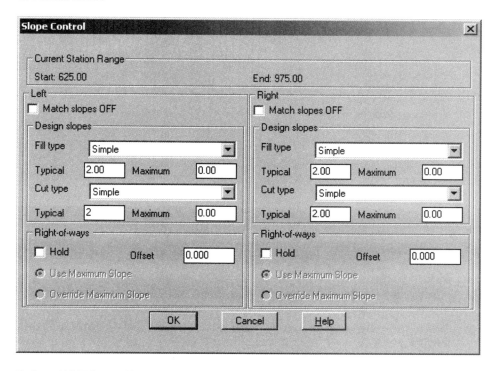

- Select "OK" to allow Land Desktop to process your sections. If you have any errors ignore them for the time being. You have successfully bound the "Road" template to your centerline alignment about the FG reference point for the station range 625-end of alignment.

- Select "Cross Sections→View/Edit Sections" to view your sections. Use the "View" option to toggle off the "Top Surface" and "ROW Lines" and to change the "Vertical Factor" to 2. Notice that the template changes at the stations that you specified. Notice that the slopes change for the stations 625+.

- Run through the section plot settings and set the options as shown in the graphic on the following page.

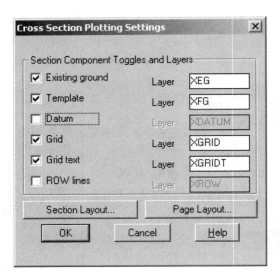

- Plot your new sections at the coordinates (18000,6000). Compare the sections to the profile to ensure that they correlate.

- Zoom extents and while on the layer "C-ANNO-VPRT" create a 40 scale landscape floating viewport which displays your cross sections.

- Save and exit your drawing. You have successfully completed this assignment.

Civil Design Control-Templates

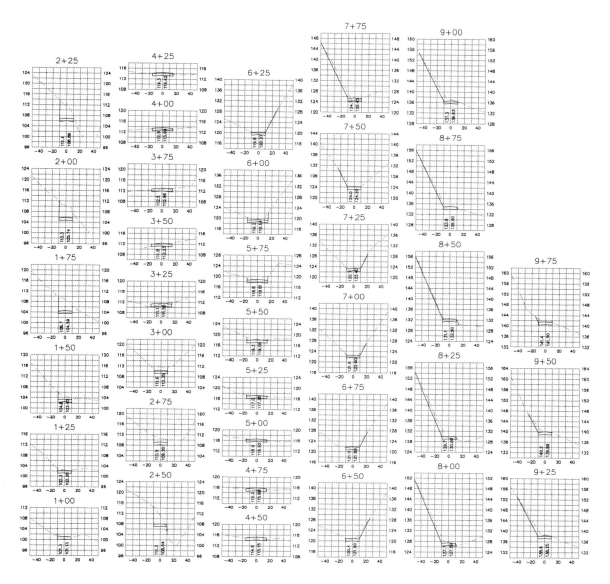

Reference Figure: Assignment 23

NOTES:

Assignment #24
Civil Design Control-Alignments

Recommended Assignments Prior to Working this Assignment:

Assignments 1-7, 12-14, 17-23

Required Assignments Prior to Working this Assignment:

Assignments 18-19, 22-23

Goals and Objectives

      In the last assignment we learned how design control is used for attaching templates to alignments about the FG Reference point. In the previous assignment, for each range of stations that users specified, the template had a constant width. The road width correlated with the size the template was drawn and defined.

      How might users address the issue of a road having a tapered width? One time consuming alternative might be to create multiple templates simulating this taper. In this assignment we will analyze a more sophisticated means of approaching this issue as we expand our knowledge of "Design Control".

      Design control gives users the ability to connect a templates control points to pre-defined alignments. These alignments are generally referred to as **hinge alignments**. In this assignment we will create two new alignments which parallel our centerline alignment and add turnouts on the left and right sides of the road. We will then use design control to attach these alignments to our template.

Exercise Instructions

- Logon to your workstation and begin a session of Land Desktop. If Land Desktop has been configured on your workstation to display the "Start Up" dialogue box, then cancel this feature so that the AutoCAD model space environment is displayed. If launching Land Desktop brings you directly into the AutoCAD model space environment, then Land Desktop has been configured to begin without the startup dialogue. If the AutoCAD Map "Task Pane" is shown, then close this dialogue box as well. Users that have the ability to customize their profiles can turn these features off permanently by following the procedure described in Assignment #1.

- Use the Land Desktop project manager (Projects→Project Manager) to copy "Assign23-##" to a new project named "Assign24-##" where the "##" reflects the number which was assigned to you on the first day of class. If the number assigned to you is a single digit (for example 4 as opposed to 14), then enter a zero preceding the single digit (such as 04). Refer to Assignment 4 if you can't remember how this was accomplished.

- In the "Copy Project To" "Name:" location type "Assign24-##" where the "##" reflects the number which was assigned to you on the first day of class. In the "Description:" and "Keywords:" areas, enter the text "Civil Design Control - Alignments". Choose "OK" and Land Desktop will ask you if you want to reassociate the copied drawing files so that the copied drawings reference the copied project. Choose "Yes" and then "Close".

- We now have a drawing file named "Assign23-##" in our "Assign24-##" project. It is important that we delete this file before proceeding so that there is no confusion as to which file is the correct "Assign23-##" drawing file. Use Windows Explorer or some other file manager to delete the "Assign23-##.dwg" drawing which resides in your "Assign24-##" project.

- Begin a new drawing by selecting "File→New" to bring up the "New Drawing" dialogue box. Type the "Name:" Assign24-## where the "##" reflects the number which was assigned to you on the first day of class.

- We are presently working with Project: "Assign24-##" where the "##" reflects the number which was assigned to you on the first day of class. Make sure that the "Project Name:" displays the correct project, select the "Acad.dwt" template and then "OK" to generate a new drawing in the "DWG" folder of the current project. After selecting the "OK" button, you may be prompted to save changes to the previous drawing session. Since there were no objects in the previous drawing session worth saving, choose not to save changes.

- You will be prompted with the "Load Settings" dialogue box which displays a list of drawing setups to choose from. These setups are saved back to the path determined by the Network Administrator who installed the software. The default path is to the local machine and is displayed above. We will set up our parameters manually, so select "Next" to set up your project with the following parameters:

    Units Area
        Linear Units = Feet
        Angle Units = Degrees
        Angle Display Style = Bearings
        Display Precision Linear = 2
        Display Precision Elevation = 2
        Display Precision Coordinate = 5
        Display Precision Angular = 4
        "Next"

Scale Area
>Horizontal = 40
>Vertical = 10
>Paper Size = 24 x 36(D)
>"Next"

Zone Area
>"Next"

Orientation Area
>"Next"

Text Style Area
>Leroy.stp
>L100

- Then select "Finish" and a screen will display providing the user with a summary of the settings chosen. Review the settings and select "OK". Although users have the ability to save settings for retrieval with future projects, assignments in this text require that you set up the parameters manually for practice in each project.

- You will not be prompted with the "Create Point Database" dialogue box, because point information was copied from the "Assign23-##" project.

- Create a few new layers with the following parameters:

Name	Color	Linetype
C-ANNO-VPRT	White	Continuous
C-ROAD-CTLN	Green	Center
C-ROAD-HPLT	Yellow	Continuous
C-ROAD-HPRT	Yellow	Continuous
XEG	Green	Dashed2
XFG	Blue	Continuous
XGRID	245	Continuous

- While in the Civil Design module, use the command "Cross Sections→Set Template Path" to set the path to the "Tplate" directory for the current project.

- Select "Alignments→Set Current Alignment→Enter" and select your centerline alignment.

- Select "Cross Sections→Design Control→Reset Section Control". This will sever any templates bound to your current alignment.

- While on the layer "C-ROAD-CTLN" select "Alignments→Import" to import your centerline alignment.

- Create a polyline from the segments and station your alignment as shown in the figure at the end of this chapter.

- Offset the polyline 10ft in each direction to simulate the roadway hinge alignments and change the layers of each hinge alignment to their respective layers. Proceed to construct the turnouts at stations 2+00 and 7+00 as shown in the figure at the end of this chapter.

- Use the alignment command to define each of the hinge alignments as "HPLT" and "HPRT" respectively.

- Make sure that your centerline alignment is current and select "Cross Sections→View/Edit Sections" to view your sections. Take a quick look at a few of them. Notice that the sections only show the existing ground and the background grid. The finished ground is not shown even though templates were defined in the last lesson.

- Specifying the templates used along our alignment will be our next task. Select "Cross Sections→Design Control→Edit Design Control". The routine will prompt you for a start station and end station. The default will be the start and end of your alignment. Accept these and choose "Template Control" to select the 10ft wide "Road" template that you defined in a previous lesson. Set up your slopes as you did in the last lesson with typical slopes equal to 2:1. Select "OK" to allow Land Desktop to process your sections. If you have any errors ignore them for the time being. You have successfully bound the "Road" template to your centerline alignment about the FG reference point for the specified station range.

- Select "Cross Sections→View/Edit Sections" to view your sections. Use the "View" option to toggle off the "Top Surface" and "ROW Lines" and to change the "Vertical Factor" to 4. Notice that the template does not stretch at stations 2+00 or 7+00. Exit the section editor.

- Select "Cross Sections→Design Control→Edit Design Control". The routine will once again prompt you for a start station and end station. The default will be the start and end of your alignment. Accept these and choose the "Attach alignments" option. For the "Left→One→Enter" option select the alignment that you defined as "HPLT". For the "Right→One→Enter" option select the alignment that you defined as "HPRT". Select "OK" to allow Land Desktop to process your sections. If you have any errors ignore them for the time being.

- Select "Cross Sections→View/Edit Sections" to view your sections. Notice that the template now expands at stations 2+00 and 7+00.

- Now once again select "Cross Sections→Design Control→Reset Section Control" to clear the design control parameters.

- Select "Cross Sections→View/Edit Sections" to view your sections. Take a quick look at a few of them. Notice that the sections only show the existing ground and the background grid.

- Select "Cross Sections→Design Control→Edit Design Control". The routine will prompt you for a start station and end station. The default will be the start and end of your alignment. Accept these and choose "Template Control" to select the 15ft "Road" template that you defined in the last lesson. Set up your slopes as you did a while back with typical slopes equal to 2:1. Choose the "Attach alignments" option. For the "Left→One→Enter" option select the alignment that you defined as "HPLT". For the "Right→One→Enter" option select the alignment that you defined as "HPRT". Select "OK" to allow Land Desktop to process your sections. If you have any errors ignore them for the time being.

- Select "Cross Sections→View/Edit Sections" to view your sections. You will find that the 15ft template yields the same results as the 10ft template. Consider: why is this?

- Use the section plot routine to plot your new sections with the template at the coordinates (18000,7000).

- Zoom extents and while on the layer "C-ANNO-VPRT" create a 40 scale 36ft x 24ft viewport of the plan view so that only the horizontal alignments show up in your viewport. The viewport is to have the landscape orientation and the lower left hand corner of the viewport must be at the origin (0,0).

- Save and exit your drawing. You have successfully completed this assignment.

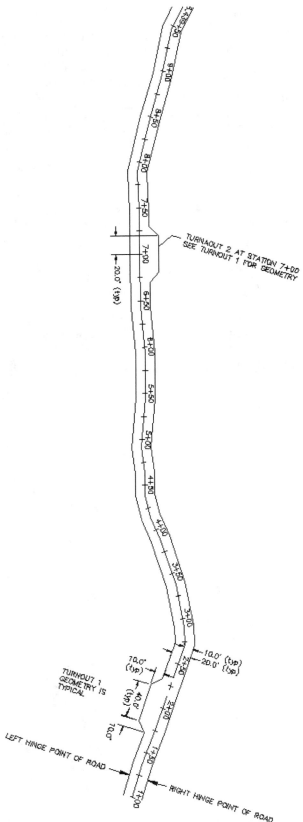

Reference Figure: Assignment 24

> Assignment #25
> ## *Civil Design Control-Profiles*

Recommended Assignments Prior to Working this Assignment:

Assignments 1-7, 12-14, 17-24

Required Assignments Prior to Working this Assignment:

Assignments 18-19, 22-24

Goals and Objectives

On occasion, a user might want to dictate template control at the hinge alignments in the vertical direction. For instance, suppose that we had a situation where the current proposed road joined an existing road at a 90 degree intersection with the existing road sloping downhill at 10%.

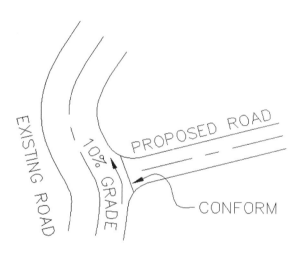

In order for our proposed road to conform with the existing road, the cross slope of our proposed road would have to be 10% at the conform. We surely don't want the entire road to have a cross slope of 10%, so we will utilize Land Desktop to smooth out the transition between a typical cross section of proposed road and the sloped existing road. Through design control, elevations may be assigned to the template at the hinge alignment locations using profiles. As we will learn in this lesson, design control gives users the ability to connect a template's control points to pre-defined profiles. We will create two new profiles to dictate template hinge elevations which parallel our centerline alignment and attach the profiles using design control.

Introduction to Land Desktop 2007

Exercise Instructions

- Logon to your workstation and begin a session of Land Desktop. If Land Desktop has been configured on your workstation to display the "Start Up" dialogue box, then cancel this feature so that the AutoCAD model space environment is displayed. If launching Land Desktop brings you directly into the AutoCAD model space environment, then Land Desktop has been configured to begin without the startup dialogue. If the AutoCAD Map "Task Pane" is shown, then close this dialogue box as well. Users that have the ability to customize their profiles can turn these features off permanently by following the procedure described in Assignment #1.

- Use the Land Desktop project manager (Projects→Project Manager) to copy "Assign24-##" to a new project named "Assign25-##" where the "##" reflects the number which was assigned to you on the first day of class. If the number assigned to you is a single digit (for example 4 as opposed to 14), then enter a zero preceding the single digit (such as 04). Refer to Assignment 4 if you can't remember how this was accomplished.

- In the "Copy Project To" "Name:" location type "Assign25-##" where the "##" reflects the number which was assigned to you on the first day of class. In the "Description:" and "Keywords:" areas, enter the text "Civil Design Control - Profiles". Choose "OK" and Land Desktop will ask you if you want to reassociate the copied drawing files so that the copied drawings reference the copied project. Choose "Yes" and then "Close".

- We now have a drawing file named "Assign24-##" in our "Assign25-##" project. It is important that we delete this file before proceeding so that there is no confusion as to which file is the correct "Assign24-##" drawing file. Use Windows Explorer or some other file manager to delete the "Assign24-##.dwg" drawing which resides in your "Assign25-##" project.

- Begin a new drawing by selecting "File→New" to bring up the "New Drawing" dialogue box. Type the "Name:" "Assign25-##" where the "##" reflects the number which was assigned to you on the first day of class.

- We are presently working with Project: "Assign25-##". Make sure that the "Project Name:" displays the correct project, select the "Acad.dwt" template and then "OK" to generate a new drawing in the "DWG" folder of the current project. After selecting the "OK" button, you may be prompted to save changes to the previous drawing session. Since there were no objects in the previous drawing session worth saving, choose not to save changes.

- You will be prompted with the "Load Settings" dialogue box which displays a list of drawing setups to choose from. These setups are saved back to the path determined by the Network Administrator who installed the software. The default path is to the local machine and is displayed above. We will set up our parameters manually, so select "Next" to set up your project with the following parameters:

    Units Area
        Linear Units = Feet
        Angle Units = Degrees
        Angle Display Style = Bearings
        Display Precision Linear = 2
        Display Precision Elevation = 2
        Display Precision Coordinate = 5
        Display Precision Angular = 4
        "Next"

    Scale Area
        Horizontal = 40
        Vertical = 10
        Paper Size = 24 x 36(D)
        "Next"

    Zone Area
        "Next"

    Orientation Area
        "Next"

    Text Style Area
        Leroy.stp
        L100

- Then select "Finish" and a screen will display providing the user with a summary of the settings chosen. Review the settings and select "OK". Although users have the ability to save settings for retrieval with future projects, assignments in this text require that you set up the parameters manually for practice in each project.

- You will not be prompted with the "Create Point Database" dialogue box, because point information was copied from the "Assign24-##" project.

- While in model space insert your Assign24-##.dwg drawing at the coordinates (0,0). Use a scale factor of one and a rotation angle of zero. Zoom extents and explode the drawing one time. We are mainly interested in the plan view, so erase the sections that were brought into this drawing with the last assignment insertion.

- While in the Civil Design module, use the command "Cross Sections→Set Template Path" to set the path to the "Tplate" directory for the current project.

- Set your centerline alignment current using the command "Alignments→Set Current Alignment→Enter" and selecting the appropriate alignment.

- Select "Cross Sections→Design Control→Reset Section Control". This will sever any templates bound to your current alignment.

- Create an existing ground "Full Profile" using the parameters established in Assignment 18. (Profiles must be created or redefined in the current drawing in order for Land Desktop to recognize the profile in the current drawings. If Land Desktop does not recognize the profile in the current drawing then users will not be able to bring in their transition alignments.) Make sure that your profile displays the existing ground and the finished ground elevations for the centerline alignment.

- Use the command "Profiles→Edit Vertical Alignments→Finished Ground", pick the finished ground centerline profile and "OK" to get to the profile editor with the desired vertical alignment current. The white toolbar with the green and red squiggly lines will allow you to copy this vertical alignment to another location. Copy the "FG Center" profile to the "FG L1" and "FG R1" profiles. After copying the profiles, close the vertical alignment editor. Elect to save your profile data when prompted.

- Use the command "Profiles→(Ditches and Transitions) DT Vertical Alignments→Import" to bring your Center, Left and Right profiles in on your newly created existing ground profile. Elect to erase objects on the finished ground profile layer when prompted. You will need to run this command three times to accomplish this. Change the center profile by layer to blue, the left profile by layer to red and change the right profile by layer to green. Notice that each of the profiles comes in on top of the previous.

- Specifying the templates used along our alignment will be our next task. Select "Cross Sections→Design Control→Edit Design Control". The routine will prompt you for a start station and end station. The default will be the start and end of your alignment. Accept these and choose "Template Control" to select the 10ft "Road" template that you defined in previous assignments. Set up 2:1 typical slopes and attach your alignments. Now select the option for "Attach Profiles" and choose the "Left One" and "Right One" options respectively. Select "OK" to allow Land Desktop to process your sections. If you have any errors ignore them for the time being. You have successfully bound the "Road" template to your centerline alignment about the FG reference point for the specified station range.

- Select "Cross Sections→View/Edit Sections" to view your sections. Use the "View" option to toggle off the "Top Surface" and "ROW Lines" and to change the "Vertical Factor" to 4. You will notice that your template is not crowned. This is because your profiles at the template hinge points have the same elevation as your profile at the FG centerline.

- Suppose that our new road is leaving an existing road to the right at 90 degrees. The existing road slopes downhill at 10%. Set up the left and right profiles to simulate this scenario. This means that the crown in the road will be eliminated. The road should have a cross slope banked to the left at station 1+00 equal to 10%. Set up your left and right profiles to also yield a 2% cross slope to the left from station 3+25 through the end of the alignment while maintaining the elevations of your FG center profile. Hint: the amount you raise and lower your transition profiles is a function of the road width.

- Re-import your left and right profiles. Your profile view should now contain 4 profiles. (PEGC, PFGC, PFGL1, PFGR1)

- Run through design control and use the option to "Attach Profiles". The other options will not need to be changed or updated. "View/Edit" your sections and have a look. The road should be sloped 10% to the left at station 1+00. Subsequent sections will show the cross slope decreasing to 2% and as having a 2% cross slope to the left for the remainder of the road (with the exception of turnout locations). The turnout locations will not appear correct without adding PVI's at their start and end. This assignment does not require that this be done.

- Use the section plot routine and while showing your template import your new sections at the coordinates (18000,8000).

- Zoom extents and while on the layer "C-ANNO-VPRT" create a 40 scale landscaped 36ft x 24ft viewport of the newly plotted cross sections.

- Save and exit your drawing. You have successfully completed this assignment.

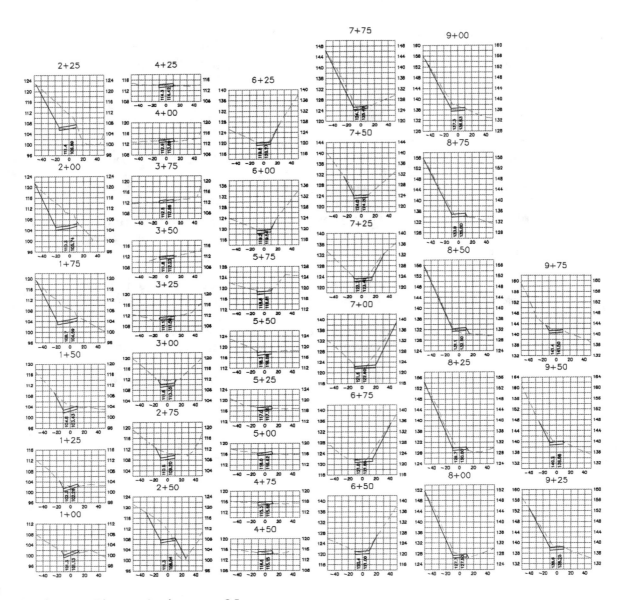

Reference Figure: Assignment 25

Assignment #26
Civil Design Volume Computation

Recommended Assignments Prior to Working this Assignment:

Assignments 1-7, 12-14, 17-25

Required Assignments Prior to Working this Assignment:

Assignments 18-19, 22-25

Goals and Objectives

You have just completed your first private roadway design as an Engineering Technician. Before submitting your improvement drawings for review, you anticipate that the local reviewing agency is going to ask for the amount of soil being moved on your site so that they can assess the project review fee. While brain storming, you seem to remember that volumes could be computed between two surfaces using the Land Desktop Terrain Menu by one of three methods. (You should review the assignment that addressed these computations if you cannot remember what they are.)

You then realize that you actually want to compute the cut and fill volume along an alignment using the Civil Design Module by average end area for the entire length of new roadway. Upon taking out your old notes, you were able to find the following:

Land Desktop volume computations using the Civil Design Module by average end area method make reference to several criteria. Before we can analyze these criteria, we have to understand how Land Desktop is obtaining the volume. The volume of a region is computed by multiplying the length of the region, its width and the height. Since the product of the width and height yields the area, it is safe to say that the volume of a region may be obtained by multiplying the area and length. "**Average end area**" method takes the average of the areas for two consecutive cross sections and multiplies by the length between them to obtain a volume.

The routine first asks the user whether or not they wish to employ curve correction. If "**Curve Correction**" is used, then Land Desktop computes the volumes in curves by using the length between the centroids of the cross sections. If curve correction is not used, then Land Desktop computes the volumes in curves by using the length between the stations (measured along the centerline).

The routine will next ask the user whether or not they want to use "**Volume Adjustment Factors**". These factors can be used to apply ratios for shrinkage or swell of material. There are times when a user might want to compute the balance based on compaction factors. These values are generally obtained from the soils engineer.

The routine finally prompts for the sample length by requesting user input for the start and end of the alignment. Once these values have been established, a table is generated summarizing various data. Items included are the cut and fill area of cross section at each station, the distance between stations, the interim volumes, the composite volumes and the mass ordinate. The "**Mass Ordinate**" is the difference between the cut and fill. Attention should be paid to the units for these values.

Exercise Instructions

- Logon to your workstation and begin a session of Land Desktop. If Land Desktop has been configured on your workstation to display the "Start Up" dialogue box, then cancel this feature so that the AutoCAD model space environment is displayed. If launching Land Desktop brings you directly into the AutoCAD model space environment, then Land Desktop has been configured to begin without the startup dialogue. If the AutoCAD Map "Task Pane" is shown, then close this dialogue box as well. Users that have the ability to customize their profiles can turn these features off permanently by following the procedure described in Assignment #1.

- Use the Land Desktop project manager (Projects→Project Manager) to copy "Assign25-##" to a new project named "Assign26-##" where the "##" reflects the number which was assigned to you on the first day of class. If the number assigned to you is a single digit (for example 4 as opposed to 14), then enter a zero preceding the single digit (such as 04). Refer to Assignment 4 if you can't remember how this was accomplished.

- In the "Copy Project To" "Name:" location type "Assign26-##" where the "##" reflects the number which was assigned to you on the first day of class. In the "Description:" and "Keywords:" areas, enter the text "Civil Design Volumes". Choose "OK" and Land Desktop will ask you if you want to reassociate the copied drawing files so that the copied drawings reference the copied project. Choose "Yes" and then "Close".

- We now have a drawing file named "Assign25-##" in our "Assign26-##" project. It is important that we delete this file before proceeding so that there is no confusion as to which file is the correct Assign25-## drawing file. Use Windows Explorer or some other file manager to delete the "Assign25-##.dwg" drawing which resides in your "Assign26-##" project.

- Begin a new drawing by selecting "File→New" to bring up the "New Drawing" dialogue box. Type the "Name:" "Assign26-##" where the "##" designates the number which was assigned to you on the first day of class.

- We are presently working with Project: "Assign26-##". Make sure that the "Project Name:" displays the correct project, select the "Acad.dwt" template and then "OK" to generate a new drawing in the "DWG" folder of the current project. After selecting the "OK" button, you may be prompted to save changes to the previous drawing session. Since there were no objects in the previous drawing session worth saving, choose not to save changes.

- You will be prompted with the "Load Settings" dialogue box which displays a list of drawing setups to choose from. These setups are saved back to the path determined by the Network Administrator who installed the software. The default path is to the local machine and is displayed above. We will set up our parameters manually, so select "Next" to set up your project with the following parameters:

    <u>Units Area</u>
    Linear Units = Feet
    Angle Units = Degrees
    Angle Display Style = Bearings
    Display Precision Linear = 2
    Display Precision Elevation = 2
    Display Precision Coordinate = 5
    Display Precision Angular = 4
    "Next"

    <u>Scale Area</u>
    Horizontal = 40
    Vertical = 10
    Paper Size = 24 x 36(D)
    "Next"

    <u>Zone Area</u>
    "Next"

    <u>Orientation Area</u>
    "Next"

    <u>Text Style Area</u>
    Leroy.stp
    L100

- Then select "Finish" and a screen will display providing the user with a summary of the settings chosen. Review the settings and select "OK". Although users have the ability to save settings for retrieval with future projects, assignments in this text require that you set up the parameters manually for practice in each project.

- You will not be prompted with the "Create Point Database" dialogue box, because point information was copied from the "Assign25-##" project.

- Create a new layer with the following parameters:

<u>Name</u>	<u>Color</u>	<u>Linetype</u>
C-ANNO-VPRT	White	Continuous

- While in the Civil Design module, use the command "Cross Sections→Set Template Path" to set the path to the "Tplate" directory for the current project.

- Select "Cross Sections→Design Control→Reset Section Control". The routine will prompt you to select an alignment. Choose the centerline alignment that you used for sampling data. This will sever any templates bound to your current alignment.

- Insert (I) assignment 25 at the coordinates (0,0) with a rotation of zero, zoom extents and explode the newly inserted drawing one time.

- Select "Cross Sections→View/Edit Sections" to view your sections. Take a quick look at a few of them. Notice that the sections only show the existing ground and the background grid. This will be an indication that your section control was reset.

- Select "Cross Sections→Surfaces→Set Current Surface" to open your existing ground surface. Select "Cross Sections→Existing Ground→Sample from surface" to sample your data with new 100ft swath widths. Set your tangents, curves and spirals to a value of 25. In the area for specifying additional sample control, choose to sample your start and end of alignment. Choose to overwrite the old section data when prompted.

- Specifying the template used along our alignment will be our next task. Select "Cross Sections→Design Control→Edit Design Control". The routine will prompt you for a start station and end station. The default will be the start and end of your alignment. Accept these and choose "Template Control" to select the "Road15" template.

    Set up your slopes as you did in previous lessons with typical simple slopes equal to 2:1.

    Choose "OK" and select the option for "Attach Alignments". Your alignments will be the left and right hinge alignments defined in previous lessons. Choose "OK" and select the option for "Attach Profiles" to attach your two profiles.

    Select "OK" to allow Land Desktop to process your sections. Now that you have sampled with 100ft swath widths, your errors should be minimal.

- Select "Cross Sections→View/Edit Sections" to view your sections. Use the "View" option to toggle off the "Top Surface" and "Point Codes" and to change the "Vertical Factor" to 4. All of the other options should be toggled on. Look at a few of your sections. Make sure that you see your existing ground and that your road template tracks along your FG profile. You should also check to make sure that your road template expands at the turnout locations.

- Select "Cross Sections→Total Volume Output→To File→Average End Area" with curve correction to compute the cut and fill volumes. Choose not to adjust the quantities with adjustment factors. Save the file to your project "Survey" directory with the name "Calc25.txt" Open up the text file using a text editor such as "Notepad" and take a look at what is being reported. You will notice that the cut is much greater than the fill. In order to keep costs down, we would generally want our cut and fill quantities to balance on site. In order to achieve this, we could elevate our FG profile or translate our horizontal alignment. It is important that you understand this concept, but do not alter your FG profile or alignments in this project.

- Here is one other thing that you should keep in mind. We sampled our centerline alignment at a 25ft interval. This is a long span in relation to the length of alignment. We need to remember that the shorter our sampling interval, the more accurate our volume computation. Select "Cross Sections→Existing Ground→Sample from Surface" to sample our data at an interval of 5ft. Maintain swath widths of 100 and choose to also sample data at the start and end of your alignments.

- Select "Cross Sections→Design Control→Reset Section Control". Then run through and set up "Design Control" just as you did earlier in this lesson.

- Select "Cross Sections→Total Volume Output→To File→Average End Area" with curve correction to compute the cut and fill volumes. Choose not to adjust the quantities with adjustment factors. Save the file to your project "Survey" directory with the name "Calc5.txt" Open up the text file using a text editor such as "Notepad" and take a look at what is being reported. You will notice that the volumes do not identically match the volumes generated in the last computation.

- Zoom extents and while on the layer "C-ANNO-VPRT" create a 40 scale 36ft x 24ft viewport of the plan view so that only the horizontal alignments show up in your viewport. The viewport is to have the landscape orientation and the lower left hand corner of the viewport must be at the origin (0,0).

- Save and exit your drawing. You have successfully completed this assignment.

```
Project: Assign26                                                              Date of Computer Run
Alignment: CTLN
                         END AREA VOLUME LISTING WITH CURVE CORRECTION
              Cut        Fill        Cut 1.0000    Fill 1.0000   Cut 1.0000   Fill 1.0000   Mass
  Station   Area (sqft)  Area (sqft) Volume (yds)  Volume (yds)  Tot Vol (yds) Tot Vol (yds) Ordinate
-------------------------------------------------------------------------------------------------
   1+00      25.70        1.70
                                        4.76          0.31          4.76         0.31         4.45
   1+05      25.70        1.70
                                        4.76          0.31          9.52         0.63         8.89
   1+10      25.70        1.70
                                        4.76          0.31         14.28         0.94        13.34
   1+15      25.70        1.70
                                        4.76          0.31         19.04         1.26        17.78
   1+20      25.70        1.70
                                        4.80          0.24         23.84         1.50        22.34
   1+25      26.17        0.89
                                        4.85          0.16         28.69         1.66        27.03
   1+30      26.17        0.89
                                        4.85          0.16         33.53         1.82        31.71
   1+35      26.17        0.89
                                        4.85          0.16         38.38         1.99        36.39
   1+40      26.17        0.89
                                        4.85          0.16         43.23         2.15        41.07
   1+45      26.17        0.89
                                        8.47          0.11         51.69         2.26        49.43
   1+50      65.28        0.25
                                       12.09          0.05         63.78         2.30        61.48
   1+55      65.28        0.25
                                       12.09          0.05         75.87         2.35        73.52
   1+60      65.28        0.25
                                       12.09          0.05         87.96         2.40        85.56
   1+65      65.28        0.25
                                       12.09          0.05        100.05         2.44        97.61
   1+70      65.28        0.25
                                       31.18          0.02        131.23         2.46       128.76
   1+75     271.47        0.00
                                       50.27          0.00        181.50         2.46       179.03
   1+80     271.47        0.00
                                       50.27          0.00        231.77         2.46       229.31
   1+85     271.47        0.00
                                       50.27          0.00        282.04         2.46       279.58
   1+90     271.47        0.00
                                       50.27          0.00        332.31         2.46       329.85
   1+95     271.47        0.00
                                       57.42          0.00        389.74         2.46       387.27
   2+00     348.71        0.00
                                       64.58          0.00        454.31         2.46       451.85
   2+05     348.71        0.00
                                       64.58          0.00        518.89         2.46       516.43
   2+10     348.71        0.00
                                       64.58          0.00        583.47         2.46       581.00
   2+15     348.71        0.00
                                       64.58          0.00        648.04         2.46       645.58
   2+20     348.71        0.00
                                       59.45          0.00        707.49         2.46       705.03
   2+25     293.33        0.00
                                       54.32          0.00        761.81         2.46       759.35
   2+30     293.33        0.00
                                       54.32          0.00        816.13         2.46       813.67
   2+35     293.33        0.00
```

Reference Figure: Assignment 26

Assignment #27
Civil Design Grading Objects

Recommended Assignments Prior to Working this Assignment:

Assignments 1-7, 12-15

Required Assignments Prior to Working this Assignment:

None

Goals and Objectives

      Your firm has been retained to prepare site grading and drainage drawings for a new custom build residence. Although the architectural drawings have not been completed, the property owner has asked that grading drawings be prepared, so that she can obtain a grading permit and have a sitework contractor begin construction this summer.

      The owner requested that the area surrounding the home appear relatively flat and conveyed that it was desirable to save all of the 14 existing oak trees which will surround the future home. Your supervisor has provided you with the preliminary architectural layout, the topographic map and asked that you prepare a conceptual grading analysis.

Exercise Instructions

- Logon to your workstation and begin a session of Land Desktop. If Land Desktop has been configured on your workstation to display the "Start Up" dialogue box, then cancel this feature so that the AutoCAD model space environment is displayed. If launching Land Desktop brings you directly into the AutoCAD model space environment, then Land Desktop has been configured to begin without the startup dialogue. If the AutoCAD Map "Task Pane" is shown, then close this dialogue box as well. Users that have the ability to customize their profiles can turn these features off permanently by following the procedure described in Assignment #1.

- Begin a new drawing by selecting "File→New". Choose to create a new project by picking the button labeled "Create Project". Select the "Prototype:" "Default (Feet)" from the drop down list. Give the project the "Name:" "Assign27-##" where the "##" reflects the number which was assigned to you on the first day of class. If the number assigned to you is a single digit (for example 4 as opposed to 14), then enter a zero preceding the single digit (such as 04). Write "Grading Objects" in the "Description:" and "Keywords:" areas. This will serve as a project summary which may be used to find or filter projects using the Project Manager.

- Select "OK" to bring you back to the "New Drawing" dialogue box and type the "Name:" Assign27-## where the "##" reflects the number which was assigned to you on the first day of class. Make sure that the "Project Name:" displays the correct project.

- Select the "Acad.dwt" template and then "OK" to generate a new drawing in the "DWG" folder of the current project. After selecting the "OK" button, you may be prompted to save changes to the previous drawing session. Since there were no objects in the previous drawing session worth saving, choose not to save changes.

- You will be prompted with the "Load Settings" dialogue box which displays a list of drawing setups to choose from. These setups are saved back to the path determined by the Network Administrator who installed the software. The default path is to the local machine and is displayed in the dialogue box. We will set up our parameters manually, so select "Next" to set up your project with the following parameters:

    Units Area
        Linear Units = Feet
        Angle Units = Degrees
        Angle Display Style = Bearings
        Display Precision Linear = 2
        Display Precision Elevation = 2
        Display Precision Coordinate = 5
        Display Precision Angular = 4
        "Next"

    Scale Area
        Horizontal = 20
        Vertical = 5

        Paper Size = 24 x 36(D)
        "Next"

    Zone Area
        "Next"

    Orientation Area
        "Next"

    Text Style Area
        Leroy.stp
        L100

- Then select "Finish" and a screen will display providing the user with a summary of the settings chosen. Review the settings and select "OK". Although users have the ability to save settings for retrieval with future projects, assignments in this text require that you set up the parameters manually for practice in each project.

- You will be prompted with the "Create Point Database" dialogue box. Accept the default settings and choosing "OK".

- Create several new layers with the following parameters:

Layer	Color	Linetype
C-ANNO-VPRT	White	Continuous
C-CONT-FGND-MAJR	Green	Continuous
C-CONT-FGND-MINR	85	Continuous

- Insert the drawing which corresponds to this lesson at the coordinate pair (0,0) with a rotation angle of 0 and a scale factor of 1. Lesson files may be downloaded from www.schroff1.com. Zoom extents and proceed to explode (X) the drawing one time.

- Using the Terrain Model Explorer, build a surface named "EG" from the existing ground contours. Refer to Assignment 15 if you can't remember how this was accomplished.

- Invoke the "Civil Design" menu (Project→Workspaces) and execute the command "Grading→Slope Grading→Grading Wizard". When prompted to select a polyline, line, arc, or grading object, select the dashed green polyline which represents the pad which we intend to grade. When asked to pick a side, pick outside of the future house, since this is the direction which we desire to grade.

- The "Footprint" dialogue box for the "Grading Wizard" will display. The "Base Elevation" is used to specify the datum elevation for the building pad and is the first vertex on the polyline representing the pad. Although the default is a flat building pad, Land desktop will allow users to grade a sloped pad. If desirable, this can be accomplished by assigning elevations to the vertices for the polyline representing the building pad using the editor in the lower portion of the dialogue box. Fill out the dialogue box as shown in the graphic on the next page and then select "Next".

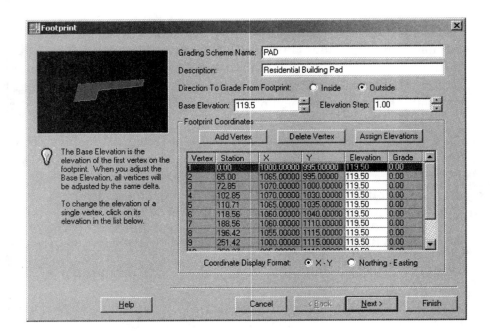

- Our Pad outline represents the object that we desire to grade from. "Targets" specify what users would like to grade to. If desirable, users can Instruct Land Desktop to grade the area East of a pad to the existing ground surface and instruct Land Desktop to grade the area West of the pad to a relative elevation. Specifying different targets along the perimeter of the pad outline can be accomplished by adding "Regions" and specifying different targets for them. We will not add Regions to our pad outline. Select the radio button adjacent to the word "Surface" and pick your "EG" surface as the target.

- Select the "Next" button to specify conform slopes. Slopes are used to specify how the area is to be graded between the pad outline and the target. If desirable, users can instruct Land Desktop to grade the area East of the Pad as having a 2:1 slope and the area West of the pad as having a 5:1 slope. This can be accomplished by segmenting the pad outline with "Tags". We will address Tags later in this lesson. Select the "Grade" radio buttons for cut and fill and specify grades of "5%" for cut and fill conforms.

- Select the "Next" button to specify how corners are to be treated. Select the radio button for "Radial Projection".

- Select the "Next" button and accept the defaults when prompted with the "Accuracy" tab.

- Select the "Next button" and accept the defaults when prompted with the "Appearance" tab. Observe the screen and select the "Finish" button. Elect to erase the existing grading object when prompted.

- Notice that the cut slope runs through one of the existing oak trees at the southeast corner of the house. We will increase the grade in this area to avoid the tree by adding Tag's to the slopes tab of the grading wizard. Select the grading object and right click to bring up a menu. Select "Grading Properties" and then the "Slopes" tab. Select the "Add Tag" button and change its value to 370 as shown in the graphic below. Also adjust the cut grade in this area to "10%".

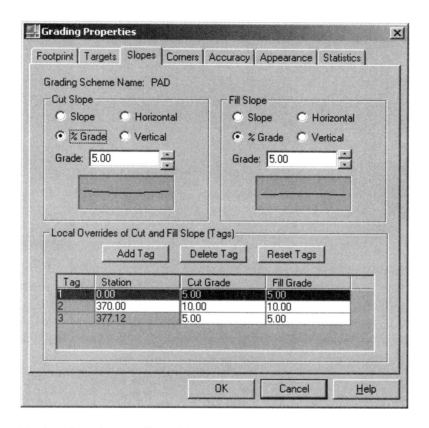

Notice that the grading object now stops short to the existing oak tree.

- Create contours which reflect the proposed grading using the menu "Grading→Slope Grading→Create Contours". Select your grading object when prompted and the "Style Manager" button to set up a contour style as was done for Assignment 14. Specify the finished ground contour layers that were set up at the beginning of this lesson and a 1ft contour interval.

- Toggle into the layout environment and create a 36ft x 24ft 20 scale landscape floating viewport in plan view on the corresponding layer. Your global linetype scale for paperspace plotting should be set to "1". The AutoCAD system variable "PSLTSCALE" should also be set to "1".

- Save and exit this drawing. You have successfully completed this assignment.

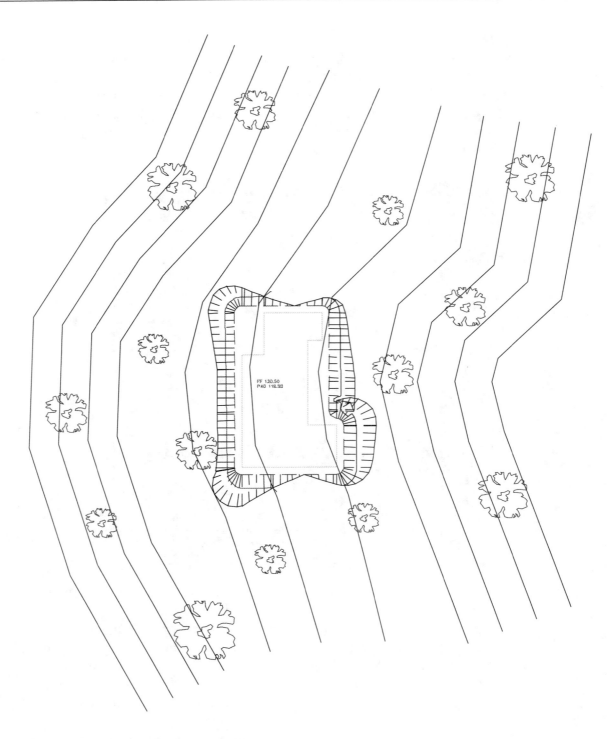

Reference Figure: Assignment 27

> Assignment #28
> ## *Civil / Survey Tool Pack*

Recommended Assignments Prior to Working this Assignment:

Assignments 1-7, 12-14

Required Assignments Prior to Working this Assignment:

Assignment 13

Goals and Objectives

This chapter is being provided to introduce a set of productivity enhancing Civil/Survey tools written by the author of this manual to work with Land Desktop 2007. This lesson more specifically targets concepts discussed in previous assignments while implementing several new productivity enhancing tools described below. The tools have been classified for the purpose of this lesson as falling under one of four categories.

1. Land Desktop Alias keys which are not available with the software and which allow users to execute several of the most widely used Land Desktop routines from the AutoCAD command line.

2. AutoCAD Alias keys which are not available with the software and which allow users to execute several existing AutoCAD routines from the AutoCAD command line.

3. A series of powerful utilities which may be run from the command line which allow users to carry out everyday tasks performed by Civil Engineers and Land Surveyors.

4. Linetypes for some of the more commonly mapped features which are not available in the software.

The Civil/Survey tool pack is not included with the purchase of this manual. Individuals who purchased this manual are simply being granted the right to use these tools free of charge for a predetermined period, at which time the tools will expire, and the lease will need to be renewed. Although users are permitted to use these tools in the workplace, the tools have not been tested in a production environment, and there is no guarantee that all of the tools will work with every Land Desktop installation. Therefore, these tools are to be used at your own risk. New tools will be written periodically and available at Schroff Development Corporation's web site for download. Users are encouraged to check the web site periodically at www.schroff1.com to download the latest productivity enhancing tools.

Introduction to Land Desktop 2007

Additional Land Desktop Alias Keys

Alias	Command	Description
BK	Break Line/Curve	Parcels->Break lines/curves command
CE	Curve from End of Object	Land Desktop's "Lines/Curves->from end of object" command
CN	Create Contours	Brings up the Dialog box to generate contours from a surface
EP	Edit Points	Opens the "Edit Points" Dialog box (must be previously initialized)
EPP	Display Properties	Edit point display properties
FF	Flip Face	Flips TIN faces (surface)
IA	Angle Inquiry	Displays relationship between two lines
IL	Line Inquiry	Lists line properties
LAP	List Available Points	Lists available point numbers.
LDD	Initialize Land Desktop	Command to initialize Land Desktop
LLD	Label Line Dynamic	Labels lines using the current label style and the dynamic label option
LND	Line By Direction	Land Desktop's "Lines/Curves->by direction" command
LNE	Line from End of Object	Land Desktop's "Lines/Curves->From End of Object" command
PG	Point Group Manager	Launches Point Group Manager
PPL	Parcels by Polyline	Defines Parcels to the database by selecting polylines
RSC	Rescale	Scales multiple objects about their insertion points
TME	Terrain Model Explorer	Opens up the Terrain Model Explorer
ZZ	Zoom to LDD Point	Allows users zoom to Land Desktop Cogo points given a point number and zoom factor

Additional AutoCAD Alias Keys

Alias	Command	Description
DMA	Dimstyle Apply	Applies current dimension style properties to a chosen dimension
DMS	Dimstyle Set	Sets a dimension style current by picking another on the screen
FR	Fillet Radius	Sets fillet radius without entering the Fillet command
JT	Justify Text	Allows users to justify text
LP	Layer Previous	Restores the previous layer configuration
PEJ	Pljoin	Joins multiple lines/LWPolylines. Note: Final product are LWPolylines (similar to multiple pedit)
PP	Pan to Point	Center drawing at current cursor location
RV	Restore View	Restores views on the fly using command line syntax
SV	Save View	Saves views on the fly using command line syntax

Civil/Survey Utilities

Alias	Command	Description
CR	CopyRotate	Copies and Rotates objects about a single base point at a user specified angle.
DE	Distance at Elevation	Calculates distance to point given two elevations and a slope.
DS	Set Snapangle to DVT	Set Crosshair angles orthoganol for the current DV Twist angle
DVL	DviewLine	Sets Dview Twist to any specified angle
ES	Elevation on Slope	Calculates the elevation given a starting elevation and a slope
HB	Hatch Back	Changes the display order of hatches by sending them to the back
HS	Horizontal Slope	Calculates slope given two elevations and a distance
IB	Image Back	Changes the display order of objects on the image layer (*RSTR*) by sending them to the back
NS	Interpolate Slope	Calculates intermitent elevations at intermitent distances given two elevations and an overall distance
OM	Multiple Offset	Offsets objects multiple times
SB	Status Bar	Turns on display of text style, dimension style, DV twist, LTS
SD	Draw Storm Drain	Draw storm drain
SF	Symbol Front	Changes the display order of objects on the symbol layer (*SYMB*) by bringing them forward
STL	Load Text Styles	Loads/redefines the most commonly used text styles
SUM	Sum	Sums the lengths of AutoCAD lines arcs and LWPolylines
TL	Change Tilemode	Toggles from Layout/Model and sets global linetype scale.
TS	Text Set	Sets the current text style by picking the text
VB	Valve Box	Draws a box having specified length and width and rotation
VLF	Viewport Lock OFF	Unlocks the display of all paperspace viewports on the current layout tab
VLO	Viewport Lock ON	Locks the display of all paperspace viewports on the current layout tab
VS	Profile Slope	Calculates slope of selected lines in profiles and cross sections given the vertical exaggeration

Linetypes

Linetype	Description
-----E-----	Electric
-----G-----	Gas
-----X-----	Fence Barb
----OH----	Overhead
----SS----	Sanitary Sewer
----->----->	Swale
-----G-----	Telephone
----UT----	Utility Trench
-----G-----	Water
----FW----	Water-Fire
----DW----	Water-Domestic
----IW----	Water-Irrigation

Exercise Instructions Part 1 - Loading the Tools

- Logon to your workstation and begin a session of Land Desktop. If Land Desktop has been configured on your workstation to display the "Start Up" dialogue box, then cancel this feature so that the AutoCAD model space environment is displayed. If launching Land Desktop brings you directly into the AutoCAD model space environment, then Land Desktop has been configured to begin without the startup dialogue. If the AutoCAD Map "Task Pane" is shown, then close this dialogue box as well. Users that have the ability to customize their profiles can turn these features off permanently by following the procedure described in Assignment #1.

- Civil/Survey tool pack files may be downloaded from www.schroff1.com. The first three categories of the Civil/Survey tool pack can be loaded into Land Desktop 2007 in one of two ways.

Method 1

Copy the files shown below into the "Support" directory of the Land Desktop Installation. The default Land Desktop support path is:

"C:\Program Files\Autodesk Land Desktop 2007\Support"

If this path is not present on your workstation, then your network administrator may have elected not to install the software with the default settings. If this is the case, then consult with your network administrator.

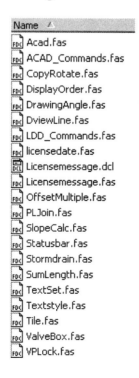

All of the files must be copied to the "Support" directory of the Land Desktop Installation, or the Tool Pack will not function appropriately.

Launch a session of Land Desktop. Invoke the "Options" dialogue box by typing "OP" at the command line. While on the "System" tab, toggle on the check box which reads "Load acad.lsp with every drawing". The tools will be ready to use the next time you launch Land Desktop.

Method 2

Launch a session of Land Desktop. Type "AP" to load applications. Select the "Contents" button under "Startup suite" and "Add" the files shown below to the startup suite by browsing to them.

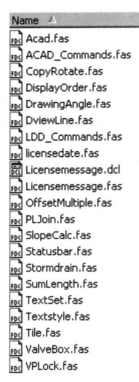

All of the files must be loaded, or the Tool Pack will not function appropriately. The tools will be ready to use the next time you open or create a drawing in Land Desktop.

- A "CEST" (Civil Engineering and Surveying Technology) pull down menu has been written for use with these utilities and may be loaded as follows:

Launch the Autodesk Custom User Interface "Tools→Customize→Interface" or by simply typing "CUI" at the command line. While on the "Transfer" tab, select the manila folder in the upper right corner of the right pane to open the customization file "CEST.cui" located with the other Assignment 28 Tool Pack files. Expand the "Menus" group in the right hand pane to view the

"CEST" menu. Expand the "Menus" group in the left hand pane to view the "Land" menu. Drag the "CEST" menu from the right pane to the left pane so that it is copied between "Utilities" and "Help". Select the apply button below to load the menu and OK to exit. The menu is not required to run the Civil/Survey tool pack since all of the tool pack commands were designed to be run from a command line.

Exercise Instructions Part 2 - Using the Tools

- Use the Land Desktop project manager (Projects→Project Manager) to copy "Assign13-##" to a new project named "Assign28-##" where the "##" reflects the number which was assigned to you on the first day of class. If the number assigned to you is a single digit (for example 4 as opposed to 14), then enter a zero preceding the single digit (such as 04). Refer to Assignment 4 if you can't remember how this was accomplished.

- In the "Copy Project To" "Name:" location type "Assign28-##" where the "##" reflects the number which was assigned to you on the first day of class. In the "Description:" and "Keywords:" areas, enter the text "CEST Tool Pack". Choose "OK" and Land Desktop will ask you if you want to reassociate the copied drawing files so that the copied drawings reference the copied project. Choose "Yes" and then "Close".

- We now have a drawing file named "Assign13-##" in our "Assign28-##" project. It is important that we delete this file before proceeding so that there is no confusion as to which file is the correct "Assign13-##" drawing file. Use Windows Explorer or some other file manager to delete the "Assign13-##.dwg" drawing which resides in your "Assign28-##" project.

- Begin a new drawing by selecting "File→New" to bring up the "New Drawing" dialogue box. Type the "Name:" Assign28-## where the "##" reflects the number which was assigned to you on the first day of class.

- We are presently working with Project: "Assign28-##". Make sure that the "Project Name:" displays the correct project, select the "Acad.dwt" template and then "OK" to generate a new drawing in the "DWG" folder of the current project. After selecting the "OK" button, you may be prompted to save changes to the previous drawing session. Since there were no objects in the previous drawing session worth saving, choose not to save changes.

- You will be prompted with the "Load Settings" dialogue box which displays a list of drawing setups to choose from. These setups are saved back to the path determined by the Network Administrator who installed the software. The default path is to the local machine and is displayed above. We will set up our parameters manually, so select "Next" to set up your project with the following parameters:

Units Area
    Linear Units = Feet
    Angle Units = Degrees
    Angle Display Style = Bearings
    Display Precision Linear = 2
    Display Precision Elevation = 2
    Display Precision Coordinate = 5
    Display Precision Angular = 4
    "Next"

Scale Area
    Horizontal = 10
    Vertical = 2
    Paper Size = 24 x 36(D)
    "Next"

Zone Area
    "Next"

Orientation Area
    "Next"

Text Style Area
    Leroy.stp
    L100

- Then select "Finish" and a screen will display providing the user with a summary of the settings chosen. Review the settings and select "OK". Although users have the ability to save settings for retrieval with future projects, assignments in this text require that you set up the parameters manually for practice in each project.

- You will not be prompted with the "Create Point Database" dialogue box, because point information was copied from the "Assign13-##" project.

- Insert the assignment 13 drawing file as a block at the coordinates (0,0) with a scale factor of one and "0" rotation angle. Explode the drawing one time. If you elected to have the drawing exploded using the check box in the dialogue box upon insertion, then do not explode the drawing a second time.

- Create a new layer with the following parameters:

Layer	Color	Linetype
C-ANNO-TEXT	WHITE	CONTINUOUS
C-ANNO-VPRT	White	Continuous
C-PROP-BNDY	MAGENTA	PHANTOM
C-SDRN-PIPE-INNR-EXST	245	DASHED
C-SDRN-PIPE-OUTR-EXST	RED	CONTINUOUS
C-WATR-SYMB	CYAN	CONTINUOUS
C-WATR-PATT	241	CONTINUOUS

- Type "TME" to launch the "Terrain Model Explorer" and set the "EG_NoFLTS" surface current. Exit the "Terrain Model Explorer".

- Type "CN" to invoke the "Create Contours" dialogue box. Construct contours having a minor elevation of 1ft and a major elevation of 5ft. Use a vertical scale factor of 1. Allow the contours to come in on their default layers. After the contours are created, go into the layer dialogue box and change the properties of the layers as follows:

Existing Layer Name	New Layer Name	New Color
CONT-MJR	C-CONT-MAJR	Red
CONT-MNR	C-CONT-MINR	245

- Isolate the layers "C-FLTS-FLIN" and "C-SURF-VIEW-FLTS-NONE" so that your display appears as follows:

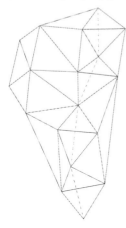

- Type "FF" and use the flip face routine to alter the orientation of the 4 surface faces which cross the flow line at locations other than vertices. This will force the faces to line up along the creek flow line and provide the same results as running a standard or proximity breakline.

- Type "CN" to invoke the "Create Contours" dialogue box. Construct contours having a minor elevation of 1ft and a major elevation of 5ft. Use a vertical scale factor of 1. Allow the contours to come in on their default layers. After the contours are created, go into the layer dialogue box and change the properties of the layers as follows:

Existing Layer Name	New Layer Name	New Color
CONT-MJR	C-CONT-MAJR-EDIT	Green
CONT-MNR	C-CONT-MINR-EDIT	95

- Isolate the two sets of contours and compare them to observe the impact of flipping surface faces.

- Turn on the points layers and type "EPP" to edit the display properties of all points so that the color of the numbers is now orange.

- Type "ZZ" and zoom to point "10" with a zoom height of "40".

- Type "EP" to edit the description of the point to read "TOP OF BANK"

- While on the layer "C-PROP-BNDY", type "LND" to create a line by direction from the starting coordinate pair (3000,25) along the bearing N30E for a distance of 150ft.

- Type "CE" and draw a counterclockwise curve from the northeast end of the previous line drawn having a 25ft radius and an interior angle of 90 degrees.

- Type "LNE" to draw a line tangent to the previously drawn arc a length of 50ft.
- Type "L" to draw a line from the end of the previous line back to the point of beginning to generate a close figure as follows:

- Type "IL" to inquire the bearing and distance of each of the lines.
    Sample: LINE DATA

    Begin . . . . . North: 25.00000   East: 3000.00000
    End . . . . . . North: 154.90381   East: 3075.00000
    Distance: 150.00   Course: N 30-00-00 E

- Type "IA" to inquire the angle between the two lines which connect at the southerly point. The acute angle is 23-11-55. The obtuse angle is 336-48-05

- Type "PEJ" and window all of the segments to join them together in the form of a polyline.

- Set your parcel settings as follows:

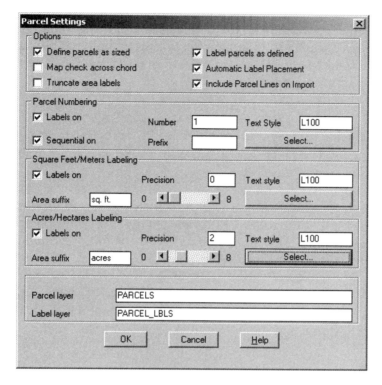

- Type "PPL" and select the polyline when prompted to define and label the parcel.

- Set up the current label style to label the direction above and distance below. While on the "C-ANNO-TEXT" layer, type "LLD" and select the polyline when prompted to label the bearings and distances.

- Type "LAP" to list the available point numbers. They should be 19+.

- Type "PG" to enter the point group manager and create a point group for all of the flowline points.

- Type "PP" and right click in the center of the recently defined parcel. This command will center the screen. Right click several times and observe how the screen is centered about your mouse clicks. When you are finished experimenting make sure your parcel is centered on the screen.

- Type "SV" then press the enter key. When prompted for a name type "P" for parcel and then enter again. A view has just been saved.

- Type "ZE" and enter to zoom extents. Now type "RV" enter "P" enter to restore your parcel view. The "SV" and "RV" commands are handy when toggling between plan, profiles and cross sections within a single drawing.

- Type "ST" to launch the text style dialogue box. Observe the text styles that are present in the drawing. Exit the text style dialogue box and type "STL" to load text styles. Type "ST" to launch the text style dialogue box and once again observe the text styles that are present in the drawing. Pay particular attention to the current text style.

- Type "TS" enter and pick one of the bearings along your parcel. You just changed the current text style to the style used for labeling your parcel. Type "ST" to launch the text style dialogue box and have a look for yourself.

- While on the layer "C-WATR-SYMB", Type "VB" to generate a valve box having a height of "5" and a length of "3" inserted at the coordinate pair (3000,25). Notice that the routine allows users to specify the orientation. The program is designed to draw and place utility boxes in topographic maps when the survey crew surveyed the center of the box. Use an angle of "0".

- Isolate the layers "C-WATR-PATT" and "C-WATR-SYMB". While on the layer "C-WATR-PATT" fill in the water box with a solid hatch. Turn on all of your layers. Notice that the parcel lines are behind the hatch.

- Type "HB" for hatch back and observe what happens.

- Type "DR" for draw order and select the water box outline. Choose to send it to the back.

- Although users would not ordinarily intentionally send symbol layers behind other objects, it often happens inadvertently. Type "SF" to bring objects on symbols layers to the front.

- Observe the entries in the status bar below the AutoCAD environment.

| 3004.79, 25.48, 0.00 | SNAP | GRID | ORTHO | POLAR | OSNAP | OTRACK | LWT | MODEL | 1 : 1445.79 |

Type "SB" and observe the changes to the status bar. The current text style, dimension style, linescale and dveiw twist angle have all been added to the status bar. Type "SB" again to restore the status bar to its original setting.

- Draw a line from the coordinate pair (2500,75) to (2600,80). Type "OM" to offset the line multiple times 50ft in the north direction. When using this command, the software will not prompt for an offset side of the line. Negative numbers offset in the southwest direction and positive offsets are to be typed for the northeast direction. For additional offsets hold down the "Enter" key. When you want to exit the command simply type "N" to exit.

- Type "SD" to initiate the create storm drain command. Type a diameter of "24" for 24 inches. We have only set up layers for the existing storm drain, so when prompted, type "E" for existing. When prompted, select one of the lines which you drew and offset. Zoom in and look at the storm drain line that was created. This program works on lines, arcs and polylines.

- Type "SUM" and type "All" when prompted to select objects. Press the F2 key on the keyboard to bring up the history. Notice that the command adds up and reports the lengths of all lines, arcs and polylines selected. This program is handy when doing material takeoffs. Press F2 again to minimize the history window.

- Zoom in on the southwest corner of storm drain which you recently created. If you knew that the existing pipe invert at the southwest end had an elevation of 100.50ft, and that the storm drain ran uphill to the northeast having a slope of 1.5%, you could calculate the elevation with a calculator if you knew the length of the storm drain. The command "ES" calculates elevations given an elevation in and a slope. Type "ES" enter. When prompted for the Elevation Begin, enter "100.50". When prompted for the Grade in % enter "1.5". When prompted for the distance, users can type in a distance, or pick two points in the model space environment. Pick the ends of the storm drain using your object snaps and the elevation at the northeast end will be reported at the command line.

- The "HS" command may be used to calculate the slope given the elevations at two ends of a utility and the distance between them. Type "HS". When prompted for the Elevation Begin, type "100.5". When prompted for the Elevation End, type "102". When prompted for a distance, users can type a distance or pick the endpoints of the storm drain in plan view.

- The "NS" interpolates elevations given two elevations and the distance between the elevations. Type "NS". When prompted for the Elevation Begin, type "100.5". When prompted for the Elevation End, type "102". When prompted for the distance between elevations, pick the two ends of the storm drain. When asked for the distance to the interpolation point, simply pick the first point along the storm drain and then a point along the storm drain to obtain the elevation.

- Type "TL" to toggle into the paperspace environment. The "TL" (Tile) command changes back and forth from paperspace to model space and sets your variables "Psltscale" and "Ltscale" appropriately each time users toggle back and forth between environments.

- While on the "C-ANNO-VPRT" layer, create a landscape 36ft x 24ft viewport. While in the viewport, restore the parcel view which was saved as "P". This next tool allows users to implement a dview twist by merely selecting two points on any object. Type "DVL" for Dview line. When prompted, pick two points along the East edge of your parcel. The objects in the viewport have been graphically rotated to your preferred orientation. Type "DS" to automatically set your snap angle orthogonal for any dview twist orientation.

- Type "VLO" to lock the display of all viewports on a layout. "VLF" can be used to unlock the display of all viewports within a layout.

- You are encouraged to explore the other utilities and linetypes in the Civil/Survey tool pack, but they are not required to complete this lesson. Save and exit your drawing. You have successfully completed this assignment.

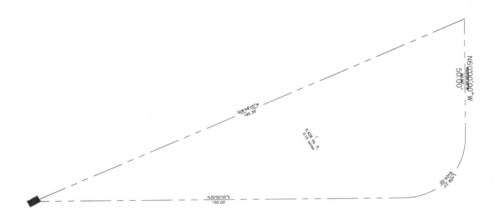

Reference Figure: Assignment 28

NOTES:

NOTES:

NOTES:

NOTES: